W9-ABZ-334

The Nature of Nurture

Sage Series on Individual Differences and Development

Robert Plomin, *Series Editor*

The purpose of the Sage Series on Individual Differences and Development is to provide a forum for a new wave of research that focuses on individual differences in behavioral development. A powerful theory of development must be able to explain individual differences, rather than just average developmental trends, if for no other reason than that large differences among individuals exist for all aspects of development. Variance—the very standard deviation—represents a major part of the phenomenon to be explained. There are three other reasons for studying individual differences in development: First, developmental issues of greatest relevance to society are issues of individual differences. Second, descriptions and explanations of normative aspects of development bear no necessary relationship to those of individual differences in development. Third, questions concerning the processes underlying individual differences in development are more easily answered than questions concerning the origins of normative aspects of development.

Editorial Board

Dr. Paul B. Baltes
Director, Max Planck Institute for Human Development and Education

Dr. Dante Cicchetti
Director, Mt. Hope Family Center, University of Rochester

Dr. E. Mavis Heatherington
Professor of Psychology, University of Virginia

Dr. Carroll E. Izard
Professor of Psychology, University of Delaware

Dr. Robert B. McCall
Director, Office of Child Development, University of Pittsburgh

Michael Rutter
Professor of Child and Adolescent Psychiatry, Institute of Psychiatry, London, England

Dr. Richard Snow
Professor of Education and Psychology, Stanford University

Dr. Stephen J. Suomi
Chief, Laboratory of Comparative Ethology, National Institute of Child Health and Human Development

Dr. Elizabeth J. Susman
Professor of Nursing and Human Development, The Pennsylvania State University

Books in This Series

Volume 1 HIGH SCHOOL UNDERACHIEVERS: What Do They Achieve as Adults?
           Robert B. McCall, Cynthia Evahn, and Lynn Kratzer
Volume 2 GENES AND ENVIRONMENT IN PERSONALITY DEVELOPMENT
           John C. Loehlin
Volume 3 THE NATURE OF NURTURE
           Theodore D. Wachs

The Nature of Nurture

Theodore D. Wachs

**Individual
Differences
and
Development
Series
VOLUME 3**

SAGE Publications
International Educational and Professional Publisher
Newbury Park   London   New Delhi

Copyright © 1992 by Sage Publications, Inc.

All rights reserved. No part of this book may be reproduced or utilized in any form or by any means, electronic or mechanical, including photocopying, recording, or by any information storage and retrieval system, without permission in writing from the publisher.

For information address:

SAGE Publications, Inc.
2455 Teller Road
Newbury Park, California 91320

SAGE Publications Ltd.
6 Bonhill Street
London EC2A 4PU
United Kingdom

SAGE Publications India Pvt. Ltd.
M-32 Market
Greater Kailash I
New Delhi 110 048 India

Printed in the United States of America

Library of Congress Cataloging-in-Publication Data

Wachs, Theodore D., 1941-
    The nature of nurture / Theodore D. Wachs.
        p.   cm. — (Sage series on individual differences and development; vol. 3)
    Includes bibliographical references (p.   ) and index.
    ISBN 0-8039-4374-1. — ISBN 0-8039-4375-X (pbk.)
    1. Nature and nurture.   I. Title.   II. Series.
BF341.W33   1992
155.9—dc20
                                                                    92-18495
                                                                         CIP

92 93 94 95 10 9 8 7 6 5 4 3 2 1

Sage Production Editor: Judith L. Hunter

BF
341
.W33
1992

260/3420

Contents

List of Tables and Figures   vii

Series Editor's Preface   viii

Preface   x

1. The Study of the Environment in
   Historical Perspective   1

2. Measuring the Environment   10

3. Specific Environmental Influences   26

4. The Structure of the Environment   39

5. Operation of the Environment Across Time   60

6. Specificity of Environmental Action   76

7. Organism-Environment Covariance   92

8. Organism-Environment Interaction   113

9. The Nature of Nurture:
   Implications and Applications   140

References   160

Index   189

About the Author   191

This book is dedicated to Benjamin and Naomi, who felt they deserved some recognition for teaching their father everything he ever knew about child rearing.

Tables and Figures

Figure 4.1 The structure of the environment   41

Figure 6.1 Patterns illustrating global and specific forms of environmental action   79

Figure 7.1 Deterministic (a) versus probabilistic (b) models of passive organism-enviroment covariance   95

Table 8.1   Inconsistent findings on organism-environment interaction   116-118

Figure 8.1 Variables influencing amount of time infant is in an active alert state   139

Figure 9.1 The environmental system   143

Series Editor's Preface

Nature (genes) and nuture (environment) are the two great sources of individual differences in development as well as a focus of this series. Contrary to the received view among developmentalists, it may be easier to study genes than environment. It could even be said that we know more than we think about genes and perhaps that we know less than we think about environment.

At the foundation of genetic influence is the triplet code of DNA. The triplet code refers to a sequence of three nucleotide bases that encodes 1 of 20 amino acids. These amino acids are combined to produce and regulate the production of all that we are. Variations in these DNA sequences, mutations, are the raw material for evolution and the source of genetic variation among individuals. Some variations such as Huntington's disease kill us, other variations have only small effects or no effect at all. Minor effects of DNA variations can add up and interact across many genes, producing major differences among children during development. The effects of such genetic variations on observed differences among children can be investigated because relatives differ in the extent of their relatedness—from identical twins who are, from a genetic point of view, clones of one another, to first-degree relatives such as siblings who are correlated 50% genetically, to genetically unrelated "relatives" such as children adopted into the same home. The power of the techniques of molecular genetics promises the dawn of an exciting new age of research in which DNA variation among individuals can be studied directly rather than inferred from familial relatedness.

There is no equivalent of the triplet code for the environment. That is, there is no fundamental, discrete unit of the environment that can be studied as the units add up and interact during development. This is why the topic is so difficult and why we may know less than we think we know. (We think we know more than we do because, unlike molecular genetics, the environment seems deceptively simple, something with which we are familiar on a day-to-day basis.) This is also why Ted Wachs's book is so important and timely. His goal is to systematize what we know about the nature of environmental influences upon development. His message is that the environment is not simple—it is a dynamic, multidimensional system whose effects are mediated by the individual.

The title of the book is a tease because the book is about the environment, not about the hyphen separating nature and nurture in the phrase *nature-nurture*. Nonetheless, more than any other environmental researcher, Ted has always kept one eye on nature as he studies nurture, which he somehow manages to do without going cross-eyed. A very real link between nature and nurture can be seen in this book in his insistence that the effects of the environment are mediated by the individual, a view that has led to his work on temperament and interactions between the organism and environment. Moreover, although this book does not focus on the integration of nature and nurture, it will promote this integration, largely launched by Ted Wachs, because it clarifies and systematizes what is known about environmental influences in development.

I have belabored this issue because my interest lies in the developmental interface between nature and nurture. For the vast majority of developmentalists who care about nurture regardless of nature, this is the book they have been waiting for since Ted Wachs's 1982 book, which was what they were waiting for since the classic 1961 book by Ted Wachs's mentor, J. McV. Hunt.

ROBERT PLOMIN

Preface

While the concept of ontogeny recapitulating phylogeny has long been relegated to the realms of scientific history, in a curious way the concept and contents of this volume mirror my own development as a researcher. That is, the ideas contained in this volume come from my own interactions with colleagues and mentors over the years. My interest in early experience as a field of study came from my own early (scientific) experience during a year spent as a research associate with the late J. McV. Hunt. From Joe I learned not only about methods and models involved in the study of early experience but also about the importance of looking at determinants of development in a broad, probabilistic sense. My ideas about the nature of environmental influences have been continually broadened by ongoing communications with a number of colleagues including Bob Bradley, Urie Bronfenbrenner, and Allen Gottfried. From a summer spent at the Behavior Genetics Institute, University of Colorado, under the tutelage of John DeFries and Gerald McLearn, I learned something about the role of genes in behavior. During a sabbatical spent with Sandra Scarr at the Institute of Child Development, University of Minnesota, I began to deal with the question of how genes and environment act together. From repeated conversations with Emmy Werner and Ernesto Pollitt, I learned not only about the relevance of nutrition to development but also that nutritional influences interact with the health status of the child and the larger cultural context within which children live. A sabbatical spent with Jack Bates at Indiana University helped refine some of my earlier ideas about the need to look at individual child characteristics, such as temperament, both as determinants and as mediators of developmental determinants.

The above individual phylogeny can be seen in the organization of this volume. The primary goal of this volume is to systematize what we know about the nature of environmental influences upon development (nurture). We start out with experience, historically (Chapter 1), methodologically (Chapter 2), and as a determinant of development (Chapter 3). In Chapter 4 we go beyond the immediate environment to look at the overall context within which the child functions and the implications of this context for understanding both developmental variability and the role of the immediate environment. Chapter 5 brings in temporal elements, showing how the environment operates across time as well as in space. Chapter 6 builds on the two previous chapters, illustrating how the existence of a multidimensional environment means that we can no longer understand the environment in a dichotomous (good-bad) fashion. Chapter 7 again goes beyond the environment, showing how the psychosocial rearing environment of the child covaries with a variety of other determinants, including genes, nutritional status, and individual child characteristics. Chapter 8 illustrates how the impact of the environment upon development will be moderated by these other determinants. In Chapter 9 I attempt to integrate the preceding chapters through the concept of the environmental system, with special emphasis on what the concept of the environmental system means for future environmental research and theory as well as for application of environmental interventions to prevent or solve real world problems.

After writing initial drafts of certain chapters, I sent these draft copies out to colleagues with special expertise in the areas covered by these chapters. Each of them took the time to provide thoughtful, incisive reviews that did much to illustrate the ambiguities in my own thinking and writing. I take full responsibility for the ambiguities that remain. For their help special thanks are offered to Marc Bornstein (Chapter 5); Jay Belsky (Chapter 6); Emmy Werner and Larry Harper (Chapter 7); Michael Rutter (Chapter 8); and Robert Bradley (Chapter 9).

I also wish to thank the series editor, Robert Plomin, not only for his gentle yet detailed comments on each chapter of this volume but also for our discussions on issues of nature and nurture. Even when we have disagreed, I have learned much. Finally, special thanks go to Cris Pecknold, who typed and retyped all too many drafts of this volume.

The Study of the
Environment in
Historical Perspective

The Prescientific Era

In discussing the evolution of ideas concerning the effect of experience upon development, it is perhaps not surprising that many major review articles start with events occurring either at the end of the nineteenth century or at the beginning of the current century (e.g., Bronfenbrenner & Crouter, 1983; Thompson & Grusec, 1970; Wohlwill & Heft, 1987; Yarrow, Rubenstein, & Pedersen, 1975). This emphasis on the recent past is not surprising, if one looks at the evolution of ideas on the role of experience. As detailed by Hunt (1961) in his classic volume *Intelligence and Experience*, from the 1700s to the start of this century, the predominant model of developmental influences was predeterminism—the concept that all developmental change (both physical and behavioral) was biologically and maturationally determined. Within a predeterministic framework, environment could have little influence on behavior or development. Prior to predeterminism the prevailing doctrine was preformationism, wherein all anatomical structures (and presumably behaviors) existed physically from the moment of conception (Hunt, 1961). Under predeterminism the development of the child could be viewed as akin to the growth of a plant—a little sunshine and a little water was all the environmental contribution that was necessary to promote what was inherent in the organism. It was only at the start of the current century that a predeterministic model of development was dethroned (at least in the biological sciences) and replaced by a more complex model, involving both biological and environmental influences upon development. Given this history it is therefore not surprising that many authors prefer to focus on recent rather than earlier ideas about the nature of environmental influences upon development.

This focus on twentieth-century ideas of environmental influences, while logical, does, however, leave out a history of ideas that is at least as long as predeterminism/preformationism, even if these ideas were not given much weight in either philosophy or science. Certainly, the idea of development not being predetermined has a long history. In the Bible (Genesis, 30:37-39, King James) Jacob is described as having caused spots to appear on newborn livestock in Laban's flock by having them conceived during the time their parents were located near trees with spots on them—perhaps one of the earliest descriptions of prenatal environmental influences. The Greek philosopher Plato, while in many respects a predeterminist (Plomin, DeFries, & McClearn, 1980), did emphasize the role of education in molding character. In his writings Plato not only noted the possibility of early environmental influences—"the beginning in every task is the chief thing, especially for any creature that is young and tender. For it is then that it is best molded and takes the impression that one wishes to stamp upon it" (The Republic, Book II, p. 177)—but also came up with a program for appropriate child rearing experiences:

> What I assert is that every man who is going to be good at any pursuit must practice that special pursuit from infancy, by using all the implements of his pursuit both in his play and in his work. For example, the man who is to make a good builder must play at

building toy houses, and to make a good farmer he must play at tilling land; and those who are rearing them must provide each child with toy tools modeled on real ones. Besides this, they ought to have elementary instruction in all the necessary subjects—the carpenter, for instance, being taught in play the use of rule and measure, the soldier taught riding or some similar accomplishment. So, by means of their games, we should endeavor to turn the tastes and desires of the children in the direction of that object which forms their ultimate goal. First and foremost, education, we say consists in that right nurture which most strongly draws the soul of the child when at play to a love for that pursuit of which, when he becomes a man, he must possess a perfect mastery. (*The Laws*, Book I, pp. 63-65)

Similarly, as pointed out by Fowler (1983), although Confucius said little about environmental influences per se, within his philosophical system nobility and virtue were not seen as predetermined but as traits that could be developed through environmental influences such as education. This belief in education was, in part, the basis of the examination system developed in China, which provided a means through which poor but intelligent individuals could rise above their hereditary status.

While these early ideas may be viewed as the forerunners of modern theories of environmental influences, for the most part these ideas were not linked together in a coherent doctrine. Indeed in many cases the idea of environmental influences was encompassed in predeterministic models, as in the case of advocating strict child rearing as a means of breaking a child's will so as to avoid the predetermined tendency of children to sin (Fowler, 1983). It was not until the work of John Locke, and his 1690 *Essay Concerning Human Understanding,* that a systematic theory of environmental influences became a viable alternative to predeterminism (Hunt, 1979).[1] Locke clearly emphasized the role of experience for the development of the mind: "Let us then suppose the mind to be, as we say, white paper, void of all characters, without any ideas; how comes it to be furnished? . . . Whence has it all the materials of reason and knowledge? . . . To this I answer, in one word, from experience; in all that our knowledge is founded and from that it ultimately derives itself" (Yolton, 1977, p. 130). Locke also specified the mechanisms through which this development occurs, namely, direct sensory input—*sensation*—and the operation of the mind upon the information it receives—*reflection*: "These two, I say, viz. external material things, as the objects of sensation; and the operations of our own minds within as the objects of reflection; are to me the only originals from whence all our ideas take their beginnings" (Yolton, 1977, p. 131).

While the strength of the predeterministic doctrine can most clearly be seen in the realm of philosophy, there is also a history of quasi experiments designed to illustrate predeterministic influences upon development. For example, Plomin et al. (1980) cite a fascinating example of an attempt by a stepbrother of Leonardo da Vinci to produce another Leonardo, through his marrying a woman similar in characteristics to Leonardo's mother.

In contrast, there were few "experiments" on the role of environmental influences upon development prior to the start of the twentieth century. One of the few examples, even though designed within a predeterministic framework, was the attempt by the medieval emperor Frederick II to determine the original language of children. As recounted by the historian Salimbene de Adam:

> He wanted to discover what language a child would use when he grows up if he never heard anyone speak. Therefore, he placed some infants in the care of wet nurses commanding them to bathe and suckle the children, but by no means ever to speak to or fondle them. For he wanted to discover whether they would speak Hebrew, the first language, or Greek, Latin or Arabic or the language of their parents. But he labored in vain because all of the infants died. For they cannot live without the praise, fondling, playfulness and happy expressions of their nurses. (Baird, Baglivi, & Kane, 1986, p. 510)

Clearly, this is one of the first recorded "studies" on maternal deprivation influences.

Other more humane experiments were based on attempts by parents to produce offspring with superior intelligence through the use of early educational enrichment. Prime examples of this type of case study rearing experiment include James Mills's attempt to teach his son Latin and Greek at the age of 3, and the work of Carl Witte, who provided an intensive program of early educational enrichment for his child starting at birth (Fowler, 1983). In both cases, whether through good heredity, good environment, or some fortunate combination of both, the offspring did, in fact, turn out to be unusually brilliant.

Even with these early attempts, it is traditional to date the beginning of *systematic* research on environmental influences to Galton's nature-nurture studies in the 1880s (Hunt, 1961). An argument can be made for an earlier date, however, based on the influence of Locke. Specifically Locke was a major influence on the French philosopher Condillac (Boring, 1957), whose writings in turn influenced the French physician Jean-Marc Itard. In 1799 a young adolescent boy was found wandering at the outskirts of the forest in which he had apparently lived for a number of years, purportedly having been reared by wolves. Leading scholars felt that the child was an incurable idiot. Influenced by both Locke and Condillac, Itard attempted to civilize the child through a detailed program of sensory training (Itard, 1962). While Itard was never able to teach the child (whom he named Victor) to speak, Victor had after five years of training acquired a variety of social skills as well as some expertise in reading and writing. Itard's work, in turn, was a major influence upon the French educator Seguin, who in turn was an influence on Maria Montessori, one of the seminal figures in early education (Fowler, 1983).

Thus, at least within the field of education, we can see an environmental intervention research tradition that clearly predates the later work of Galton. Why this research tradition did not spread more widely outside of education may be due, in part, to the influence of G. Stanley Hall. It was Hall who introduced the concept of predeterminism to the newly emerging behavioral sciences at just about the same time that the biological sciences were dropping predeterminism in favor of a more multidetermined model. Given the influence of Hall as an educator, the transmission of his ideas through students such as Arnold Gesell no doubt contributed to lowered interest in developing systematic research on environmental influences (Hunt, 1961). This delay was only temporary, however, and by the early part of the twentieth century a viable research program on the study of environmental influences was beginning to emerge.

Twentieth-Century Study of Environmental Influences

Bronfenbrenner and Crouter (1983) have suggested that the evolution of research and theory on environmental influences in the current century falls into three distinct phases. *Phase I* (the prototypic stage), which Bronfenbrenner and Crouter date from approximately 1870 to 1930, was partly focused on the question of nature versus nurture, as seen in comparisons of identical versus fraternal twins or in studies of adopted children. Phase I also was characterized by research and theory based on the study of social address models. Social address models are based on comparing the development of children from different sociodemographic groups without directly assessing the actual environments of individual children placed in these groups. Even though environment was not directly assessed, differences between children in different groups were assumed to reflect differences in experience. Elsewhere (Wachs, 1983) I have argued that what is common to Phase I research and theory is the question: "Does the environment have an influence upon development?" Although Phase I research was fraught with methodological problems, the overall conclusion, as summarized in reviews (e.g., Hunt, 1961; Thompson & Grusec, 1970), is an affirmative answer to the question of whether the environment has an influence upon development.

Phase II environmental research and theory began in the 1930s and continued into the 1950s (Bronfenbrenner & Crouter, 1983). This phase was characterized by research looking at the influence of aspects of the environment that existing theories saw as critical for subsequent development. Among the theories that were used to guide research in this second phase were the classical psychoanalytic theory of Freud, the neuropsychological theory of Hebb, the sociological-anthropological theories of researchers such as Dollard, Vygotsky, and Whiting and Child (though Vygotsky's influence upon North American and Western European research occurred well after 1950; Belmont & Freeseman, 1988), and the ethological tradition of Lorenz (Bronfenbrenner & Crouter, 1983; Hunt, 1961, 1979). From each of these theories came specific hypotheses about which aspects of the environment ought to be related to development. For example, from psychoanalytic theory, dimensions such as early stress and parental rejection were seen as prime targets for environmental research. From ethology came an emphasis on the importance of early attachment relations, while Hebb stressed the importance of repeated visual and motoric stimulation. What was common to the diverse body of Phase II research was a focus on the question of what *specific aspects of the environment* are relevant for development (Wachs, 1983). That is, these theories basically accepted the idea that environment was relevant and began to delineate what was meant by environment. Obviously not all studies during the era of Phase II research were concerned with delineating the environment. Many Phase I studies continued to be done. For example, the classic Iowa child rearing studies, which demonstrated that major gains in cognitive performance could occur when retarded infants were moved from an unstimulating to a highly stimulating institutional environment (Skeels & Dye, 1939), clearly fit the model for Phase I studies. For the most part, however, during Phase II the emphasis was upon documenting the influence of purportedly critical characteristics of the environment.[2] As documented in reviews (Hunt, 1979; Wachs & Gruen, 1982), for the most part Phase II research has been relatively successful in identifying those aspects of the environment that are particularly salient for various domains of infant and child development. The third and current phase of environmental research and theory in this century, starting sometime after 1950 (Bronfenbrenner & Crouter, 1983), appears to be driven by two assumptions: (a) that the environment is not homogeneous in nature but is divided into different levels and even into different subunits within a given level and (b) that the influence of the environment upon development will be mediated by a variety of nonenvironmental factors, including but not limited to age and individual characteristics of the child (Wachs, 1986). Thus, in

Phase III research and theory, we find an increasing emphasis upon the simultaneous study of multiple units of the environment (e.g., Crockenberg, 1981; Elder, Downey, & Cross, 1986; Wachs & Chan, 1986). Other Phase III studies look at the mediation of environment-development relations by nonenvironmental factors, such as temperament (e.g., Gandour, 1989), age (e.g., Rutter, 1981a), or sex of the child (e.g., Zaslow, 1989).

Common to all of these distinct Phase III studies is an emphasis upon what I have called the *process* aspect of environmental theory and research (Wachs, 1983). Process is defined by two domains: first, by more precise mapping of the various dimensions of the environment and their interrelationships and, second, by an emphasis on understanding the means by which variability in the environment translates into variability in developmental outcomes. In one sense Phase III represents a shift away from studying the impact of the environment upon development toward studying the nature of environment and of environmental influences.

Obviously, even in Phase III, ongoing research is still being conducted on Phase I and Phase II questions. For example, the whole emphasis of the impact of day care upon infant development (Belsky, 1988) clearly is a Phase I question (Are there environmental influences?). Similarly we have recently obtained data indicating that certain environmental factors, such as vocal stimulation and nonverbal response to the child's vocalization, relate to variability in cognitive development in less developed countries (Egypt and Kenya) in much the same way they do in the United States and Western Europe (Wachs et al., 1992). This type of study clearly represents a Phase II question (What specific environmental influences are relevant to development?).

While there is clearly a temporal overlap between these three phases, the history of environmental research in the current century has evolved from Phase I questions (Is the environment influencing development?) to Phase II questions (What are the specific components of environment that relate to development?) to the current emphasis upon Phase III process questions. While there are ample reviews available on the current status of answers to Phase I and Phase II questions (e.g., Hunt, 1961, 1979; Thompson

& Grusec, 1970; Wachs & Gruen, 1982), there have been few systematic reviews of Phase III questions, models, and results. An emphasis upon Phase III environmental research and theory is the primary focus of this volume.

Notes

1. Fowler (1983), however, has cogently argued that the work of Comenius, although not as influential as that of Locke, should be regarded as the first systematic theory of environmental influences.

2. There were also some examples of Phase II research occurring during the Phase I period, as in Van Alstyne's (1929) pioneering research correlating various aspects of the home environment with the Kuhlman-Binet performance of 3-year-olds.

2

Measuring the Environment

As might be inferred from Chapter 1, there are a variety of means through which we can study the role of environmental influences upon development. When comparing these different procedures, a typical approach is to point out their strengths and weaknesses and then evaluate each procedure on a good-bad dimension. While this approach is important, it is not, nor should it be, the sole criterion upon which environmental assessment procedures are evaluated. Another critical dimension that also needs to be considered is the appropriateness of a specific measure for the question being asked (Wachs, 1991a). Environmental assessment procedures that have definite weaknesses may still be of use in answering certain questions, while highly desirable environmental assessment procedures may be totally inappropriate for answering other types of questions. Previous reviews have detailed the strengths and weaknesses of environmental procedures used to answer Phase I (Does the environment influence development?) and Phase II (What dimensions of the environment are relevant?) questions (Altmann, 1974; Bronfenbrenner & Crouter, 1983; Rutter, 1981b; Sackett, 1978; Wachs, 1988a, 1991a; Wachs & Gruen, 1982; Wohlwill, 1973a; Yarrow & Anderson, 1979). Therefore we will only briefly summarize the major features of these types of designs. A major focus of this chapter will be on the characteristics, strengths, and weaknesses of environmental designs used to answer Phase III questions (the nature and process of environmental action).

Phase I Procedures

As described by Bronfenbrenner and Crouter (1983), the study of Phase I questions was based initially on either social address or nature-nurture type studies. The typical social address study compares children's development as a function of composite demographic characteristics, such as parental educational level, social class, or rural versus urban residence. The assumption underlying these studies is that differences between groups of children at different social addresses are due to environmental factors associated with different social addresses. The prime strength of social address studies is the ease with which data can be gathered, often allowing for large sample sizes and increased statistical power. The major weakness of the social address study is the fundamental assumption that group differences are due primarily to the environmental component of the social address rather than to other factors associated with social address, such as group differences in biomedical risk conditions or nutritional status (Powell & Grantham-McGregor, 1985).

Although not a traditional social address design, *cross-cultural studies* appear to share many of the same virtues and defects. In cross-cultural studies the assumption is made that developmental differences between children in different cultures are caused by culturally associated differences in environmental rearing conditions. Developmental differences between children of different cultures, however, also may be a function of nonrearing factors, such as culturally based gene differences (Shand & Kosawa, 1985) or differences in incidence or severity of child illness (Pollitt, 1983).

An alternative form of the social address design is epidemiological research. Of particular relevance are "experiments of nature," which involve relating developmental outcomes to specific ecological contexts encountered by children (e.g., prolonged institutionalization, changes in family structure such as divorce, changes in the quality of schools, forced migration). To the extent that these naturally occurring influences are random in terms of whom they affect, epidemiological research allows us to avoid some of the confounds associated with traditional social address research. The critical question is whether the proper controls and cross-checks have been implemented to test for potential confounds in epidemiological studies (Rutter, 1981b). Potential confounds include determining whether observed results are a function of epidemiological factors or the nature of the sample being studied, or whether observed associations between epidemiological variables and outcomes are due to the influence of a third factor common to both variables (Rutter, 1981b).

The nature-nurture designs most often used to answer Phase I questions are adoption studies and twin studies. The basic logic of adoption studies is that a child who is adopted early in life receives only its genes from its biological parents and only its environment from its adoptive parents. The degree to which the child resembles the adoptive parent is viewed as an index of the degree of environmental influence. For twin studies, because monozygotic (MZ) twins share both common genes and common environments, while dizygotic (DZ) twins mainly share common environments, greater similarity between MZ and DZ twins on a trait can be taken as evidence for environmental influence upon that trait. Some studies combine the twin and adoptive methodology, looking at the similarity or dissimilarity of monozygotic twins who are reared apart.

While the logic of adoption and twin studies is fairly straightforward, drawing conclusions from these studies about environmental influences can be highly problematic. The degree to which environmental conclusions drawn from adoption studies are valid depends upon meeting a number of assumptions, some of which are mutually exclusive. For example, it is important to have children adopted as early as possible, to minimize nonadoptive environmental influences. Yet children who are adopted early are more likely to be brighter or more socially appealing than children who are adopted later, so that later adopted children may be reared longer in less adequate nonadoptive environments (Munsinger, 1975). In addition, the type of statistics used may determine whether adoption results can be interpreted as indicating genetic or environmental influences, *even when the same data are used*. As Weizmann (1971) has pointed out, correlational statistics tend to be insensitive to mean differences. This means that there could be a significant correlation between the adoptive child and biological parent's scores (indicating genetic influences) at the same time that the child's scores are well above the level of the biological parent's scores (indicating environmental influences). This possibility is not just abstract speculation (e.g., see the Minnesota Adoption Studies—Scarr & Weinberg, 1983).

For twin studies interpretation of results also depends upon a number of methodological assumptions. Chief among them is the question of whether MZ and DZ twins share the same environment, or whether the environment of MZ twins is more similar than that of DZ twins (which would tend to inflate the MZ correlation). Whether this differential similarity of environment actually exists, and what this differential similarity might mean, continues to be a topic of debate (Hoffman, 1991; McCartney, Harris, & Bernieri, 1990).

One way of dealing with the problem of potentially greater environmental similarity for MZ twins would be to look at the resemblance between MZ twins who are reared apart (combining the adoption and twin designs). Even here there is ambiguity, however. Using a more differentiated model of the environment (see Chapter 4), Bronfenbrenner (1986) has noted that, even when MZ twins are being reared in different households, they may share other aspects of the environment. Reanalyzing the data from a number of twin studies, Bronfenbrenner (1986) reports that the correlation between adoptive MZ twins living in similar communities was .86, whereas the correlation between adopted MZ twins living in different communities was only .26. Thus to obtain an adequate assessment of environmental influences from studies of

MZ twins being reared apart requires much more attention to the meaning of being reared apart than has heretofore been paid.

Overall it seems clear that nature-nurture studies can be used to assess the relevance of environmental contributions to development, as long as appropriate caution is exercised. Given the tendency of twin studies to underestimate environmental influences, these studies may be less appropriate than adoption studies, particularly adoption studies that look at mean differences in performance as well as parent-child correlations.

The final class of studies that can be used to answer Phase I questions are general intervention studies. If individual children or groups of children are randomly assigned to different treatments, then cross-group differences between children should reflect the impact of environmental influences. While general intervention studies appear to be the most powerful approach to answering Phase I questions, there are also potential methodological and conceptual problems, which may limit the conclusions about environmental influences that can be drawn from intervention studies. First there is a question of generalizability. Many general intervention projects are focused only on children who are extremely economically disadvantaged. Whether these same programs would work with children who are less disadvantaged remains an open question (Woodhead, 1988). In addition an intervention program may work in one context but may be totally irrelevant in a second context (Scarr & McCartney, 1988). In addition researchers also must be careful about whether groups that are assigned to different treatments are, in fact, truly random. Preexisting differences between groups in terms of either history or selection factors will be confounded with treatment. Under certain circumstances this may lead to an overestimation of treatment (environmental) effects (Madden, O'Hara, & Levenstein, 1984).

Finally, many intervention programs assume that all children will benefit more or less equally from a given intervention. In fact, some children may be very much affected by a given intervention while others may receive little benefit (Wachs, 1988a). Thus, in addition to using mean differences between groups as a criterion for assessing the impact of environmental interventions, it is critical also to look at *variability within intervention groups* to determine what proportion of children in a given treatment are actually benefiting. While general intervention studies are perhaps the most powerful way we have of demonstrating the existence of environmental influences, the relative simplicity of this approach should not blind us to problems inherent in this method as well.

Phase II Studies

Phase II studies center on the question of which environmental variables are related to variability in development. Given that a different question is being asked, certain research designs that were useful in answering Phase I questions may not be applicable for answering Phase II questions. For example, the utility of social address studies for answering questions about the salience of specific environmental parameters has been repeatedly questioned (e.g., Bronfenbrenner & Crouter, 1983; Wachs, 1983) due to the tremendous variability in environments encountered *within* a given social address. The traditional nature-nurture study, which looks at adoptive children or twins without defining the specific nature of the adoptive or twin environment, suffers from many of the same problems and thus is not particularly appropriate for answering Phase II questions. Similarly, general stimulation studies, which involve presenting target children with a complex, multidimensional package of stimulation, are rarely appropriate for answering Phase II questions. While development may be promoted by this package of stimulation experiences (Phase I question), this type of research tells us little about what specific components of the stimulation package are most salient for influencing developmental outcomes (a Phase II question).

In contrast, cross-cultural studies, while less useful in answering Phase I questions, may be quite useful under certain conditions. By looking only within a given culture, the impact of certain experiences that are *common* to all children in that culture may be overlooked. By comparing cultures where the specific experience is or is not present, it is possible to test the importance of these "invisible" variables (Bornstein, 1989a). One excellent example would be school attendance: Are age-related gains in cognitive performance due to the cumulative experience of school attendance or to maturation (Cahan & Cohen, 1989)? In industrialized countries the overwhelming majority of children go to school, whereas not all children go to school in many nonindustrialized countries. To the extent that we can control for within-culture differences on who attends school, we can use nonindustrialized countries as a setting to test whether school attendance has an influence (e.g., Stevenson, Parker, Wilkinson, Bonnevaux, & Gonzalez, 1978). Obviously these types of studies do not tell us what about the school environment influences children's development. This alternative Phase II question can be answered, however, by looking at differences between school systems on variables such as classroom size, organizational structure of the school, nature of the classroom environment, and degree of academic emphasis (Rutter, 1983a).

In terms of research procedures specifically designed to answer Phase II questions, both *interview* and *report measures* typically involve asking parents or caregivers about the types of experiences they provide to their child. The main virtue of parent report measures is their flexibility, in that parents can be questioned about the environment provided for their children across a number of content domains and contexts. Parent report measures also may be particularly useful in situations where it is necessary to document rarely occurring events, such as family trips. Parent questionnaires also can be easily and quickly administered to large numbers of parents, thus increasing the sample size and the potential power of the research design. Problems, however, also exist with these types of measures. For the most part, parents' reports about past experiences provided for their child have proven to be highly inaccurate when compared with documented evidence about what actually was happening (Yarrow, Campbell, & Burton, 1970). While concurrent reports may be more accurate, one major problem is that concurrent reports may be based not on what the parents actually are doing with their child but on what the parents feel they ought to be doing (Wachs & Gruen, 1982). Thus it is not surprising that the degree of convergence between observer ratings and parent's report may be quite modest (Kochanska, Kuczynski, & Radke-Yarrow, 1989). For these reasons the use of parent report measures about the *nature of their child's environment* appear to have limited value in terms of answering Phase II questions.

Parent report measures also may be used to assess parent *attitudes,* either toward their children or about specific child rearing practices. Parent attitude measures are easy to administer. While parental attitudes are related to parental behavior, for the most part the magnitude of correlations between attitudes and behaviors tends to be quite modest (Kochanska, 1990; Miller, 1988). Thus, if the focus of the research question is on child rearing practices, then the utility of attitude measures is again quite limited. If the research focus is on attitudes, however, and not environment per se, then the low attitude-behavior correlations may be of less concern. Even when used just to measure attitudes, however, the utility of these types of measures has been called into question. For example, in a recent review on the psychometric adequacy of parent attitude measures, Holden and Edwards (1989) report major measurement and conceptual problems with all existing parent attitude measures.

A third type of parent report measure involves parental perceptions about the extent and impact of nonfamily environmental factors, such as stress conditions or social and emotional support from family and nonfamily members. For these types of measures an argument can be made that what is critical is the caregiver's perception of the existence of social support or stress. That is, if the caregiver feels that support is available, the caregiver may well behave as if support is available, whether or not it is. Particularly if researchers are interested in links between different levels of the environment, this type of information may be particularly useful as an efficient, low cost approach to measuring the nature of environmental variables outside of the home.

One alternative way of dealing with some of the problems inherent in parent report measures would be through the use of *child report measures.* Particularly for older children, it has been argued that what is important is the child's perception of specific experiences rather than the parent's interpretation or the observer's rating (Kagan, 1967). The use of child's report as an index for assessing the importance of specific aspects of the environment has a number of advantages. Children can be asked about their perceptions of their environment across a wide range of family and nonfamily experiences (e.g., school), thus maximizing representativeness. Because child perceptions can be obtained from questionnaires, this also means the possibility of a relatively large sample size. Available evidence, however, suggests that reports of children younger than 10 years of age tend not to be reliable (Edelbrook, Costello, Dulcan, Kalas, & Conover, 1985). As a result this class of measures may be useful primarily for older children and adolescents.

The logic of *manipulative studies* is to vary one or more aspects of the child's environment to determine the impact of this variation upon the child's behavior or development. Manipulations may be directed toward individual children, groups of children (classroom), parents, or even physical features of the environment. Manipulations may even be epidemiological in nature, taking advantage of naturally occurring environmental variation. One example of an epidemiological manipulation study is seen in the work of Elder and his colleagues (e.g., Elder, Van Nguyen, & Caspi, 1985) on the developmental implications of growing up during the Great Depression of the 1930s. Another is seen in studies investigating how variation in naturally occurring noise (children who live near airport runways or whose nursery schools are close to subway tracks) influences children's development (e.g., Cohen, Evans, Stokols, & Krantz, 1986).

While both individual and group manipulative studies would seem to be an ideal way of assessing the salience of specific environmental variables, in reality great caution must be exercised in interpreting the results of these studies. For example, one of the supposed virtues of group treatment studies is that these studies typically involve large numbers of children, which suggests that these studies should have greater statistical power. As Cronbach and Snow (1977) have noted, however, the correct unit of analysis in group treatment studies should be the group rather than the individual, which may severely limit "sample" sizes. Even more critically, in many group treatment studies researchers often make the assumption that children given treatments with different labels are encountering functionally different environments, while children given treatments with similar labels are in equivalent environments. Treatment labels, however, may be, at best, only an imperfect description of the specific environmental parameters to which children are actually exposed (Carbo, 1983; Cronbach & Snow, 1977; Keogh, 1986a). To obtain the necessary preciseness, detailed observation of specific features of group treatments are critical.

Typically individual treatment studies have tighter control over what is being manipulated. Sample sizes tend to be smaller, however, which may limit power. More critical is the problem of whether what is being manipulated is an *accurate reflection* of what the child normally encounters in his or her natural environment—the problem of representativeness or ecological validity (Bronfenbrenner, 1977). As noted by Wohlwill (1973a) and McCall (1977), while individual manipulation may well cause a developmental change, this tells us nothing about whether these manipulations actually exist in the real world and, if they do, whether they cause the same type of change. To the extent that individual or group manipulative studies are an accurate reflection of what the child encounters in the real world, these studies may be useful in answering Phase II questions about the salience of specific environmental influences.

Direct observation procedures encompass the recording of naturally occurring physical or social aspects of the child's environment and relating these measures to variability in children's development. By directly measuring what is happening in the child's natural habitat, the chances of obtaining accurate and representative measures of the child's environment are maximized, as long as certain methodological techniques are implemented.[1] While the presence of an observer often will cause initial distortions in caregiver behaviors (Zegiob, Arnold, & Forehand, 1975), with *repeated observations* caregiver behaviors tend to return to their normal pattern (Wachs & Gruen, 1982). Given that the pattern of caregiver/child transactions may be different when other family members are present (Clarke-Stewart, 1978; Wachs, 1986), it is important to do observations at *different times* of the day rather than at just a single point in time. Because environmental assessments obtained from single short-term observations are unlikely to be stable across time, it is important to use either *repeated short-term observations* (Wachs, 1987a) or *single long-term observations* across a period of at least several hours (Lytton, 1980). On the negative side, if obtaining representative, stable environmental assessments means repeated direct observations across multiple contexts and times, this type of design is highly time consuming. As a result many direct observation studies have relatively small sample sizes. Small sample size studies are more likely to have low statistical power, which may mean a reduction in our ability to detect the impact of specific environmental parameters. Suggestions on how to deal with this problem are provided in Chapter 9.

While Phase II observational studies often involve observation in naturally occurring contexts (home, day-care center, classroom), there are studies that use observation of caregiver-child relations in a nonnaturalistic setting, such as a laboratory playroom. While these types of studies allow more precise control of extraneous variables, there remains the question of whether what is being observed is an accurate reflection of what the child normally encounters in more "real world" settings (Belsky, 1980; Novak, Olley, & Kearney, 1980). Discrepancies between laboratory and real world observation may be due to different patterns of behavior by children in familiar versus unfamiliar situations (Plunkett, Berlin, Dedrick, Dichtelmiller, & Meisels, 1990; Walden & Baxter, 1989). The degree of correspondence between home and laboratory observations also tends to be lower when highly overt laboratory observations are used (Field & Ignatoff, 1981) or when mothers are given little guidance as to how to structure their behaviors (O'Brien, Johnson, & Anderson-Goetz, 1989). To some extent it may be possible to overcome these problems through use of repeated measurements of caregiver-child transactions in more naturalistic laboratory settings. An excellent example of this type of study is seen in the work of Kochanska et al. (1989), who designed a laboratory to look like a small apartment (kitchen, bathroom, playroom) and observed depressed mothers and their toddlers interacting in this setting over several occasions.

Phase III Studies

The distinction between Phase II and Phase III studies can be seen in the argument by Plomin and Bergeman (1991) that some of the variability in development, which typically is attributed to specific environmental dimensions, may be associated with genetic factors that covary with these environmental dimensions. Whether variability in children's development is a function of the covariance between genes and environment is clearly a Phase III process question. To answer this Phase III question accurately, however, it is necessary to have Phase II evidence defining what are salient environmental parameters. Phase II studies define what are the salient environmental parameters; Phase III studies tell us how these parameters influence variability in development.

One important question raised in many Phase III studies is how different *levels* of the environment interrelate to influence development. For the most part research on this question requires integration across the different methods typically used in either Phase I or Phase II studies. One common approach is to interrelate questionnaire measures of parental perceptions of social support or stress, either with observational measures or with parent or child perception measures, to determine whether higher order environmental factors like support or stress directly influence development or whether their influence is due to their impact upon the child's rearing environment. For example, Cotterell (1986) has reported that paternal absence from the home due to work pressures (parental report) influences both maternal reports of her parent rearing attitudes (questionnaires) and the quality of the mother's interaction with her preschool child (naturalistic observation). Other Phase III studies relate naturally occurring environmental events or social addresses to more specific child rearing dimensions and thus to development. For example, the impact upon adolescent adjustment of the interrelation between an epidemiological event (severe economic depression) and child rearing practices (maternal interview) has been documented by Elder et al. (1985). Similarly, combining Phase I and II procedures, Iverson and Walberg (1982) report that the relation between parent reports of their rearing practices and children's academic achievement was influenced by the SES level of the family.

It should be noted that Phase III studies of this type are not restricted solely to investigating relations between higher order environmental variables and the home environment of the child. Studies also can look at interrelations among higher order environmental variables, as these affect development. For example, the detrimental impact of urbanization (an epidemiological factor) upon family relations and child development may be mediated by cultural practices involving the degree of contact the father traditionally has with the child in specific societies (Davidson, 1980). Similarly the presence of social support may serve to buffer children against the detrimental impact of epidemiological factors such as exposure to armed conflict (Klingman & Wiesner, 1982).

The utility of these cross-level designs can be increased substantially by adding in a longitudinal dimension. An excellent example of this approach, which has been dubbed the study of *causal chain mechanisms,* is seen in the research by Rutter (1985a; Rutter & Pickles, 1991) investigating the adult adjustment of females who were reared in disadvantaged backgrounds. Demonstrating the operation of cross-level environmental causal chains, Rutter has reported that psychosocial competence in adulthood was found to be associated with a combination of social address factors in *childhood* (being reared in an institution), along with higher order environmental factors in *adolescence* (self-reports about the nature of the child's school experience) and in *adulthood* (quality of marital relationships). A similar example is seen in the long-term follow-up by Werner and Smith (1982) of extremely high risk children, about 10% of whom managed to develop into well-adjusted adults. Among the infancy factors that led to later differential adjustment of very high risk children were *social address* (age of opposite sex parent), *family structure* (number of sibs), and *quality of psychosocial interactions* (moderate attention given to the child). In childhood and adolescence additional contributions were made by *family structure* (availability of sib or caregiver whom the child could use as a resource), *family climate* (household rules and family cohesiveness), and *epidemiological factors* (number of stressful events encountered).

What all of the above studies indicate is how the methods described earlier for Phase I and Phase II questions can be integrated to arrive at answers for Phase III questions, relating the structure of the environment to children's behavior and development. Obviously, the same methodological strengths and weaknesses associated with each method apply regardless of whether these methods are used in Phase I, Phase II, or Phase III designs. Further, there are additional methodological problems encountered in causal chain designs. These include avoiding selective loss of subjects across time and minimizing the impact of repeated testing (Wohlwill, 1973a). The payoff, in terms of understanding how multiple levels of the environment interact and influence development across time, however, seems more than obvious from the examples given above. In cases where the obstacles encountered in pure cross-level/cross-time causal chain designs are too formidable, various "shortcut" designs offer some of the same advantages while minimizing some of the problems. One example would be the convergence method, which consists of a series of overlapping short-term longitudinal samples (for a detailed description of these and other cross-time designs, see Wohlwill, 1973a).

A second major area of focus for Phase III studies involves describing the *process*, wherein variability in environment relates to variability in development. Phase I and II research designs also can be tailored to answer these types of specific theory-driven questions. For example, Plomin (Plomin & Daniels, 1987; Rowe & Plomin, 1981) has argued that nature-nurture designs can be used to partial environmental variance into those components that are shared by individuals within families (and make individuals more alike) versus those environmental components that are unique to individual children and lead to different developmental outcomes. While the question of whether environmental factors act to make individuals different or similar is a Phase III question, the use of traditional nature-nurture designs to answer this question is still somewhat controversial (see Chapter 6 for further discussion of this controversy).

Naturalistic observation approaches also may be useful in answering Phase III questions, to the extent that naturalistic observations are used to test *specific models of environmental action*. For example, Bradley, Caldwell, and Rock (1988) used a combination of longitudinal naturalistic observation and parent reports to compare the validity of three models of environmental action: (a) whether relations between early environment and later development were primarily a function of early experience, regardless of what environmental influences came later; (b) whether relations were primarily a function of later environment, regardless of the nature of the early environment; (c) whether relations were due to the cumulative impact of stable environmental influences across time. Similarly both Crockenberg (1987) and Wachs (1987b) have used naturalistic observation procedures to test whether individual child characteristics mediate relations between environment and development.

Observations in structured settings or manipulative studies, while less useful for assessing Phase II questions, may be particularly appropriate for the study of Phase III process questions. Observations done in laboratory settings not only allow researchers to ask process questions but also have the advantage of allowing greater control over extraneous environmental variables than is possible with observations done in naturalistic settings. For example, using repeated observations in a laboratory setting with toddlers between 14 and 22 months, Heckhausen (1987a) was able to test a model of environmental action, which predicted that mothers adapt their instructional style to their infants' perceived level of competence rather than to their infants' chronological age.

Both individual and group environmental manipulation studies also can be used to test specific process models of environmental action. For example, Parpal and Maccoby (1985) trained a group of mothers to allow their preschool age children to direct the mothers' behavior in joint play situations. By comparing the compliance behavior of children whose mothers were trained versus children whose mothers were not trained, Parpal and Maccoby were able to test whether compliance behavior in preschool children was due to parent-child reciprocity, parent reinforcement patterns, or social deprivation of the child. At the level of group manipulations there is a rich literature testing the validity of aptitude by treatment models of environmental action. Typically aptitude by treatment designs involve training teachers or pupils in specific educational procedures (the environmental intervention) and then testing whether the impact of these procedures is mediated by individual difference factors such as pupil ability, anxiety, or attitudes (e.g., Corno, Mitman, & Hedges, 1981; Webb, 1989). As with Phase II studies, manipulations do not have to be programmed by the experimenter. Naturally occurring epidemiological manipulations also may be useful in testing process models of environmental action. One example is the study by Lumley, Ables, Melamed, Pistone, and Johnson (1990), who investigated the mediating effects of temperament upon children's coping behavior in response to a naturally occurring environmental stress, namely, short-term hospitalization.

What seems clear is that new techniques are not necessarily needed to investigate Phase III questions. Adaptation of traditional methods is more than appropriate, as long as these traditional methods are used to test specific models of environmental action or to answer specific questions about the nature of environmental influences. Obviously, in using traditional designs to answer Phase III questions, the researcher must be well aware of the strengths and limitations of each design. As long as these considerations are kept in mind, it seems clear that there are no right or wrong ways to investigate the salience, characteristics, or nature of environmental influences. Rather there are a variety of methods that are more or less appropriate for answering specific questions about the nature of the environment and its influence upon development. What is most critical is for the researcher to specify exactly what phase question is being tested, prior to determining which environmental assessment procedure is most appropriate.

Note

1. Although the issues involved are too technical to deal with in a general review chapter, it is important to note that the type of observational procedures chosen may influence the representativeness of information obtained. For example, recent evidence demonstrates that time-sampling procedures may result in less accurate measurement of typical mother-infant behaviors than continuous sampling procedures (Mann, Have, Plunkett, & Meisels, 1991).

Specific Environmental
Influences

From Phase I studies we know that differences in rearing environments are related to differences in children's cognitive and social-emotional development. Evidence supporting this conclusion comes from a variety of sources, including both environmental intervention studies (e.g., Barnard & Bee, 1983; Grantham-McGregor, Schofield, & Powell, 1987; Rauh, Achenbach, Nurcombe, Howell, & Teti, 1988) and adoption studies (e.g., Dumaret, 1985; Scarr & Weinberg, 1983). Available evidence indicates that increased environmental stimulation can even facilitate physical growth in malnourished children (Super, Herrera, & Mora, 1990). What these Phase I studies do not tell us is what *specific dimensions* of the environment account for variability in development. This chapter briefly reviews evidence from Phase II studies on this question. Obviously, given the tremendous amount of information in this area, not all potentially relevant aspects of the environment can be reviewed here. The goal of this chapter is to highlight those dimensions of the environment that have been most *consistently* related to development. Because of space limitations and the extremely large num-

ber of Phase II studies, direct discussion of specific studies is n̸
feasible. The conclusions drawn in this chapter will primarily be
based on major reviews, which summarize what we know about
the role of environmental influences. Readers interested in greater
detail about topics raised in this chapter, or in information about
a "favorite" environmental variable that is not cited, can refer to
the following sources:

(a) reviews emphasizing *cultural or societal influences* include
    Bronfenbrenner (in press), Kagitcibasi and Berry (1989), and Rogoff
    (1990);

(b) reviews emphasizing *nonfamily environmental influences* include Belsky
    (1990), Crouter and McHale (in press), Garmezy (1983), McLoyd (1990),
    and Rutter (1983a);

(c) reviews emphasizing *direct family influences* include Graham (1977),
    Gottfried (1984), Hetherington (1989), Rutter and Garmezy, (1983),
    Seginer (1983), Sigel (1985), Wachs and Gruen (1982), Wahler and
    Dumas (1989), and Wohlwill and Heft (1987).

A note of caution (caveat emptor): When reading the next three
sections, readers are urged to keep in mind that many of the
relations discussed here are not necessarily generalizable to all
developmental outcomes. Rather the impact of specific environmental predictors will vary as a function of the *overall context* (see
Chapter 4), *age of the child* (Chapter 5), *specific outcomes under study*
(Chapter 6), and *individual biological or child characteristics* (Chapters 7 and 8). Given the goals of this chapter, however, the environmental dimensions discussed here are presented as if we were
assuming that each dimension operated in isolation.

Cultural and Societal Factors

COGNITIVE FUNCTIONING

While much of the research on cultural and societal influences
involves children's social-emotional development, some evidence
is available indicating the potential relevance of different cultural

nitive development. Cultural beliefs about which
es are seen as important appear to translate into
fferences on what cognitive abilities will be stressed
r informal education of children. Put another way,
while cognitive structures or the capacity for learning may hold
across all cultures, cultural factors can determine which cognitive
control factors are critical, what is to be learned, and at what age
specific skills should be learned (Wagner, 1978). For example, in
certain African cultures social interaction skills and manual abilities are central to a definition of "intelligence." As might be
expected, child rearing patterns in these cultures emphasize the
development of social and manual abilities. Another example of
this process is seen in the development of cognitive style parameters,
such as field independence versus field dependence (whether the
person uses an internal or external frame of reference in dealing with
the environment) and psychological differentiation (the complex-
ity and integration of individual behavioral patterns). There is a
compelling body of evidence suggesting that the development of
cognitive styles is a function of both the natural ecology and the
cultural structure associated with that ecology. Specifically ecologies that lead to hunter-gatherer life-styles require independence
and self-reliance, which in turn are associated with the development of field independence and high psychological differentiation. In contrast ecological-cultural settings that predispose to
agricultural life-styles require cooperation, social sensitivity, and
the development of social controls, which in turn are associated
with the development of field dependence and lower levels of
psychological differentiation. Further, as noted in the pioneering
work of Vygotsky (1978), as cultures change new cognitive skills
may be emphasized at the expense of more traditional skills. Thus,
as traditional societies become more industrialized, requiring different cognitive skills, the relation of ecological factors to the
development of field independence and psychological differentiation seems to decrease (Witkin & Berry, 1975).

SOCIAL-EMOTIONAL DEVELOPMENT

To the extent that parental values differ across cultures and relate to rearing patterns, we can expect children in different cultures to be exposed to different types of rearing styles. These differences in rearing styles should result in different developmental patterns in areas such as self-concept, moral judgment, and the nature of social relationships. For example, in Japan cultural values emphasize the group. These values are translated into specific rearing practices such as prolonged body contact between mother and infant as well as disciplinary practices that stress empathy with others' feelings about the child's transgressions and exclusion from the family (e.g., in America "grounding" at home is viewed as a punishment; in Japan being locked out of the home would be the cultural equivalent of grounding). The end product is a child who values within-group cooperation over competition and has a high sensitivity to interpersonal cues (Weisz, Rothbaum, & Blackburn, 1984).

As another example, in many parts of Africa cultural values emphasizing obedience and respect lead to a lessening of adult attention and a greater proportion of adult commands, demands, and reprimands as children go from infancy into early and middle childhood. This pattern results in a child who, while highly active and sociable with peers, also will be highly inhibited and obedient in the presence of adults (Harkness & Super, 1982). Similarly, available evidence indicates that the level of characteristic adult aggression relates to the degree of aggression shown by children within specific cultures (Fry, 1988). Given that children may learn a culturally appropriate pattern for expressing aggression, this process allows for the maintenance of aggressive behavior over generations.

The descriptions given above reflect differences in stable, value-governed rearing patterns *across* cultures. In addition there are also within-culture influences that can potentially affect most people living in that society. Three examples, which are particularly notable in the current century, are increasing industrialization or urbanization, war or revolution, and economic disruption.

For many traditional cultures increasing urbanization and industrialization have been implicated as causing reduction in family interactions, increased intergenerational conflicts, and inadequate socialization of children (e.g., Leng & Ong, 1982; Sanda, 1982). The consequent impact on children, at least according to clinical observations, includes increased vulnerability to behavioral problems, loss of a sense of personal control, decrease in the ability to delay gratification, and increase in adolescent identity confusion (e.g., Morice, 1980; Thanaphum, 1980). There also is evidence indicating that the impact of urbanization may be mediated by the child's perception of the urban environment (Weisner, 1981) as well as by existing cultural patterns such as the degree of father-child contact (Davidson, 1980).

There is also an increasing body of evidence on the impact upon children of war, revolutions, or terrorist activities within a culture. Clearly children's ongoing behavior can be adversely affected during acute phases of fighting; in addition, the by-products of cultural violence, such as forced migration, appear to be associated with children developing a sense of learned helplessness (Werner, 1979). These conclusions, however, must be tempered by other evidence indicating that children appear to display tremendous resilience in the face of major cultural disruptions such as war (Garmezy, 1983). Indeed the direct impact of war or terrorism on children's own aggressive behavior is not yet clear, with some studies showing that children who were victims of violence grew up to be violent adults (Fields, 1979), whereas other studies find surprisingly little impact on children growing up in the midst of violence (McWhirter & Trew, 1982). In part the impact of cultural disruptions such as war appears to be mediated by a variety of factors, including level of social support, quality of parent reactions, and individual characteristics such as sex of child.

A similar pattern is seen when we consider the effects of society-wide economic disruption. The impact of economic depression has been demonstrated to influence children's development in a variety of ways. Some of the impact is negative, such as increased depression and delinquency (Lempers, Clark-Lempers, & Simons, 1989). Not all of the impact need be negative, however; higher levels of independence and responsibility also may result (Bronfenbrenner, in press). Again these differences may be due to the fact that relations between culturally wide economic disruption and children's development may be mediated by a variety of factors, including sex and age of the child.

Nonfamily Environmental Factors

COGNITIVE COMPETENCE

While there is some evidence that nonfamily factors such as social support may influence children's early cognitive development (e.g., Bee et al., 1982), for the most part research on the influence of nonfamily factors has involved the study of schooling. It seems clear that school attendance can promote the development of a variety of cognitive processes, perhaps in part because schools teach specific cognitive skills that are valued within individual cultures. (However, schools also may teach children to be comfortable with being tested or with carrying out specific adult requests even though a request may be meaningless to a child). There is, of course, tremendous variability among schools in terms of their impact on children, which raises the question of which school characteristics are most likely to promote the development of cognitive competence. Within minimal limits, for preschool as well as primary and secondary schools, physical facilities, number of books in the library, school size, and amount of money spent per pupil do not appear to relate strongly to cognitive or academic achievement. What does appear to be relevant are what might be called expectancy parameters, such as an orientation toward achievement or student perceptions of the school as cohesive, organized, and goal directed. Also relevant are organizational factors, such as regularity in requiring and marking homework and teacher time spent actually teaching and not maintaining order and discipline. Good class structure and sensitivity to students, as evidenced in feedback to students and reward for good performance, also seem to be important. Also salient for the educational performance of older children is a nonschool factor, namely, *peer group educational aspirations.* Available research suggests a modest but consistent impact of peer group aspirations, with higher peer aspirations being associated with higher achievement, particularly in urban settings (Ide, Parkerson, Haertal, & Walberg, 1981).

While it seems clear that schools can have a facilitating impact on certain dimensions of cognitive competence, a cautionary note has been sounded by clinicians in developing countries. Because of the tremendous competition for a limited number of places in higher grades, school age children in these countries may be subjected to high levels of stress and family pressure. As a result a child's failure to meet family achievement demands may have a serious negative impact upon the child (Graham & Canavan, 1982). In such situations schooling may be a mixed blessing for some children. Similarly, in situations where the classroom situation does not fit traditional cultural patterns (e.g., cultures where an older sib may be the primary caregiver rather than an adult), a traditional teacher-led instructional format may not be appropriate (Tharp, 1989).

SOCIAL-EMOTIONAL DEVELOPMENT

In addition to affecting cognitive performance, schools also can have a socializing influence on children. Available research has noted that school attendance can reduce delinquency, enhance self-esteem, and increase achievement motivation. Other studies suggest that the development of a sense of internalized responsibility is associated with schools that emphasize students' self-direction of learning experiences and that grant students more privileges and more responsibilities. In contrast schools that discourage pupil initiative and active pupil involvement in school affairs may promote a feeling of learned helplessness.

As with cognitive development, *peers* also can be a source of influence on social-emotional development. Through appropriate interaction with peers, older children can pick up such skills as perspective taking, empathy, and the ability to modulate aggression (Hartup, 1983). Thus it is not surprising that exposure to nondelinquent peers may reduce chances of further delinquency in delinquent boys (Zarb, 1978), whereas exposure to nonretarded peers can promote higher levels of social competence in retarded children (Simmeonsson, 1978).

In recent years, given the increasing number of two-wage-earner families, there has been an increasing interest in the impact of preschool experiences such as *day care* upon young children's individual and interpersonal development. Debate continues to rage upon whether day-care experiences in the first year can have an adverse impact upon children's development. Some researchers have argued that early day care may lead to insecure attachment and higher levels of aggressive behavior (e.g., Barglow, Vaughn, & Molitor, 1987; Haskins, 1985). Other researchers have argued that the strange situation paradigm, which is often used to assess attachment, may be inappropriate for day-care children, who are routinely separated from their parents; alternatively, what is being called increased aggressiveness may, in fact, actually reflect increased independence (Clarke-Stewart, 1989).

While debate about the positive versus negative impact of day care continues unabated, there is agreement on the importance of high quality day care for promoting children's subsequent individual or interpersonal development. To a great extent high quality day care is defined both by design (physical) features of the day-care environment as well as by caregiver factors. Specifically design features such as more personalized settings, child-scaled environments, well-defined landmarks, and well-organized space are among the setting features used to define high quality day care (Weinstein, 1987). Critical caregiver features include *stability* (the less turnover in caregiver staff, the higher the quality of day care), amount of *training* (the more caregiver training, the higher the quality of day care), the *caregiver to child ratio* (the fewer children per caregiver, the higher the quality of day care), and caregiver *verbal interaction, involvement,* and *responsivity* (the more verbally interactive, involved, and responsive the caregiver, the higher the quality of day care; Clarke-Stewart, 1989; Phillips, McCartney, & Scarr, 1987).

Other indirect nonfamily influences that have been shown to influence children's social-emotional development are social support and the work environment of the parents. The work environment of the parents may influence their values, which in turn may be transmitted to their children. For example, available evidence indicates that employed mothers emphasize independence training with their children more than nonemployed mothers (Hoffman, 1989). Work stresses also may adversely affect the parents' mood or behavior at home or may limit the amount of time the parent has to monitor child activities, all of which could adversely affect children's social-emotional development. Similarly a lack of social support may result in parents being less sensitive to their children, thus leading to developmental deficits.

Direct Environmental Influences

COGNITIVE COMPETENCE

A number of major reviews have confirmed that, for the early years of life, there are a variety of physical and social characteristics of the child's home environment that are particularly critical for early cognitive performance. Aspects of the physical environment that are positively related to cognitive competence include *availability* of stimuli (particularly in the first nine months of life), *variety* of stimuli (changes in available objects becoming more critical than number of available objects as the child gets older), *responsivity* of the physical environment, and *organization* or regularity of scheduling in the home. In contrast evidence consistently suggests that high levels of ambient background *noise, overcrowding,* and *"home traffic pattern"* (number of people coming and going in the home) are negatively related to cognitive performance.

Social-environmental factors in the home that are consistently and positively related to cognitive competence in the early years of life include degree of parent *involvement*, level of *tactual* or *kinesthetic* stimulation (particularly for the first six months of life), *contingent responses* to the child's *distress* (in the first year of life), and *verbal stimulation* (particularly after 12 months of age). Conditions that deny the child opportunities for locomotor exploration or direct parental *restrictions on exploration* tend to be negatively related to the development of early cognitive competence.

As children move into the preschool years, while a number of the factors noted above continue to be relevant, different family environment factors also become important for the development of cognitive competence. One example would be the use of *direct skill teaching* by parents or caregivers. The process of task-oriented dyadic interaction between caregiver and child has been described using a variety of different terms, such as *joint teaching, structuring, scaffolding,* or *guided participation.* Underlying all of these different terms is the idea that children's cognitive development is facilitated through joint, task-oriented activities with an adult, who guides a child through problem-solving experiences with tasks that the child would normally not be able to solve by him- or herself. One critical aspect of dyadic task-oriented interaction involves the *pacing* of the interaction. Optimal development occurs when the adult does not pressure the child to learn, leave the child to deal with tasks that are too difficult for the child to handle, or take over the task from the child. Rather optimal development occurs when adults adjust their level of interaction to the child's own level of competence. It is also important to note that, in many cultures, the main caregiver is a sib or peer rather than an adult. While cognitive gains are less likely to occur when the young child is interacting with same-age peers (Turkheimer, Bakeman, & Adamson, 1989), there is some evidence that interaction with older sibs or older child caregivers can have a developmentally facilitative effect upon the cognitive development of younger children (Teti, Bond, & Gibbs, 1988; Wachs et al., 1992).

For school age children and adolescents, a number of studies have consistently indicated a positive relation between parents' holding high *achievement values* and their children's actual achievement. Active parent *involvement* in the child's educational activities, parental *monitoring* of the child's activities, and parental *guidance* also have been shown to facilitate older children's cognitive development and academic achievement. Other positive influences include a *variety of intellectually stimulating materials* in the home. Certain aspects of *family structure* also seem to influence cognitive development, among them absence of the father, which has consistently been reported to have a negative impact. Family factors also may influence cognitive style parameters, with evidence indicating that, in families valuing independence of action, children tend to be field independent, whereas parents who value conformity have children who are more likely to be field dependent.

SOCIAL AND EMOTIONAL DEVELOPMENT

Across the first few years of life, parental *sensitivity* and parental *responsivity* to the infant have been demonstrated to be particularly critical for various aspects of infant and toddler development. Of particular importance is the repeatedly demonstrated relation between parental sensitivity and subsequent infant attachment behavior. The importance of sensitivity for children's development, however, may go well beyond attachment per se. Specifically sensitivity may promote secure attachment in infants, which in turn may promote greater exploration of the environment (Ainsworth & Bell, 1970), greater toddler compliance and problem-solving competence (Frankel & Bates, 1990), and more appropriate interactions with peers in middle childhood (Sroufe & Egeland, 1991). After infancy one critical environmental factor that becomes increasingly influential for children's development of impulse control is the nature of parental *control strategies.* Direct power assertive parental control strategies such as anger, physical punishment, or criticism often lead to less impulse control and less compliance. In contrast parental responsiveness to the child's control needs, as seen in the use of *reciprocal* power sharing in play (the use of give and take strategies), as well as the use of *reasoning* and *suggestions* when the parent finds it necessary to assert control, have been consistently associated with greater impulse control, compliance, and the development of self-reliance by young children (Crockenberg & Littman, 1990; Holden & West, 1989).

A similar pattern is seen in older children and adolescents. Consistently harsh parental *punishment, rejection* of the child, and *inconsistent discipline* tend to predispose to antisocial, aggressive, and delinquent behavior on the part of the child. This relationship appears to hold cross-culturally, at least to some degree. Aggressive behavior also is promoted when parents *inadequately monitor* their children's activities or when parents negatively respond to their child's activities, regardless of whether the child's behavior is positive or negative in tone. *Low involvement* with the child also can have undesirable consequences, such as a greater probability of adolescent drug use (Block, Block, & Keyes, 1988).

Available research studies also have consistently implicated *family discord* as a major etiological factor in the production of antisocial behavior, aggression, and delinquency in children. Family discord also has been implicated as a cause of various nonaggressive behavior problems in children, such as maladaptive behavior in the classroom or poor peer relations. One type of family discord, parental disagreement on disciplinary practices, has been associated with low self-esteem and poor self-control in children (Vaughn, Block, & Block, 1988), though Deal, Halverson, and Wampler (1989) have argued that parents who disagree simply may be ineffective parents in general. Family discord may cumulate in *divorce*, with available evidence suggesting a higher incidence of adjustment problems in children from divorced families, particularly if they are boys (Hetherington, 1989). On the other hand, *family cohesiveness* has been found to promote children's task persistence and their ability to cope with environmental and internal stress (Murphy & Moriarty, 1976). Similarly, adequacy of *family communication patterns* has been found to protect children who are at risk for schizophrenia due to genetic predisposition. In addition, parental responsivity and involvement have been found to promote adolescent adjustment, while parental encouragement of the child's independence produces greater striving for independence by the child.

Summary of Environmental Influences

What has been illustrated in this chapter is the variety of environmental factors that can act to influence development. For the most part these influences have been presented as if they were "main effect" in nature, namely, as if there were a *direct line* between environment and outcome. As noted earlier, in many cases the impact of a specific environmental influence will be mediated by other environmental influences or by the child's characteristics. For example, although boys in divorced families tend to be at greater risk for adjustment problems than boys in nondivorced families, many male children in divorced families do show highly positive outcomes (Hetherington, 1989). In part these differential outcomes are a function of the child's temperament, nonfamily factors such as social support, and the quality of the relation with the noncustodial parent. Thus any attempt to understand the nature of specific environmental influences must take into account other factors, such as individual characteristics and the operation of different levels of the environment. Put another way, rather than looking at environmental influences upon development in a one-to-one linear fashion, it is important to consider environmental influences as part of a system. It is this systemwide aspect of environmental influences that will be discussed in the following chapters.

4

The Structure of the
Environment

Most definitions of the environment refer to organized conditions or patterns of external stimuli that impinge upon and have the probability of influencing the individual (Bronfenbrenner & Crouter, 1983; Wachs & Gruen, 1982). If asked to give examples of "environments" that are relevant for children's development, most individuals would emphasize the child's home. Environmental research, however, illustrates the need to go beyond the home, and look at other dimensions or levels of the environment, to understand the nature of environmental influences upon development. The major focus of this chapter will be upon the nature of different levels·of the psychosocial environment. What will be demonstrated in this chapter is not only that the environment is best understood as a *multidimensional dynamic system* but also that the impact upon development of environmental components from any one level of the system may depend upon the nature of environmental components at other levels of the system. Put another way, in this chapter we will begin the transition from Phase II (aspects of the environment) to Phase III questions—in case, how do influences of a specific environmental dimension depend upon the structure of the environmental system?

Structural Models of the Psychosocial Environment

In general different disciplines have focused upon different aspects of the environment. Geographers have emphasized physical ecology, anthropologists have emphasized culture, urban planners have emphasized neighborhoods, and psychologists have emphasized parent-child relations. Each of these different aspects, which traditionally have been treated in isolation, can be seen as distinct *levels* of the environment. One of the first attempts to come up with a conceptual model that involved multiple levels of the environment is seen in the work of Lewin (1936) and his concept of the life space. Since Lewin's pioneering writing, other researchers have attempted to develop multilevel taxonomies of the environment, encompassing anywhere from 3 to more than 200 dimensions (e.g., Block & Block, 1981; Chein, 1954; Horowitz, 1987; Moos, 1973; Sells, 1963; Super, 1981; Wohlwill, 1973b).

Currently the most well-developed and influential environmental model used by developmental researchers is Bronfenbrenner's (1989) *ecological framework,* which involves a hierarchical system consisting of four levels. A diagram of Bronfenbrenner's model is seen in Figure 4.1.

At the lowest level there is the *microsystem*, which is "a pattern of activities, roles and interpersonal relations experienced by the developing person in a given face to face setting with particular physical and material features, and containing other persons with distinctive characteristics" (Bronfenbrenner, 1989, p. 227).

Encompassing the microsystem is the *mesosystem*, which "comprises the linkages and process taking place between two or more settings containing the developing person (eg., the relations between home and school, school and work place, etc.). In other words a mesosystem is a system of microsystems" (Bronfenbrenner, 1989, p. 227).

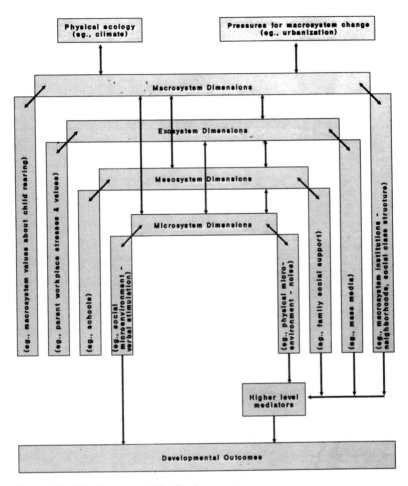

Figure 4.1. The Structure of the Environment

SOURCE: Freely adapted from various works of Bronfenbrenner.
NOTE: Double headed arrows refer to bidirectional levels of influence across the environment. Single headed arrows refer to direct or mediating influences of the environment upon development.

Encompassing the mesosystem is the *exosystem,* which "encompasses the linkage and processes taking place between two or more settings, at least one of which does not ordinarily contain the developing person, but in which events occur that influence processes within the immediate setting that does contain the person (e.g.,

for a child the relation between home and the parents' workplace . . .)"
(Bronfenbrenner, 1989, p. 227).

At the highest level is the *macrosystem*, which is defined as consisting of "the overarching pattern of micro, meso, and exosystems
characterizing a given culture, subculture or other broader social
context, with particular reference to the developmentally instigated
belief systems, resources, hazards, life styles, opportunity structures,
life course options and patterns of social interchange that are embed-
ded in each of these systems" (Bronfenbrenner, 1989, p. 228).

While it would seem that the macrosystem is equivalent to
culture, within Bronfenbrenner's system not all macrosystems
qualify as cultures or even subcultures. For example, while both
macrosystems and cultures involve shared beliefs and values,
macrosystems also can be defined on the basis of shared resources,
shared hazards, or shared historical events. Thus ethnic neighborhoods or eras (i.e., the "1960s"), while not fitting the definition of
a culture, could fit the definition of macrosystem. Bronfenbrenner's
model will serve as a framework for this chapter.

The Nature of Environmental Structure

Bronfenbrenner's model is best viewed as a *dynamic system,*
which operates across both place and time. These operations can
be characterized as follows:

1. The environment is hierarchically organized such that characteristics of the environment at one level can influence characteristics of the environment at other levels (Bronfenbrenner, 1989).[1]
This influence can occur in *two ways*:

a. Higher levels of the environment can act to *directly* influence
caregiver *values and beliefs.* For example, the macrosystem may act
to define for caregivers what are appropriate developmental tasks
or when these developmental tasks may be achieved (Kindermann
& Skinner, 1988).

b. Higher levels of the environment may influence caregiver *behaviors*, either through influencing caregiver beliefs about which child
rearing practices are effective (Luster, Rhodes, & Hass, 1989) or

through structuring how children are taught (e.g., formal schooling versus direct training; Rogoff, Mosier, Mistry, & Goncu, in press).

2. Higher level environmental processes can *mediate or modify* the pattern of relations between environment and children's development occurring at lower levels (Bronfenbrenner, 1989; Horowitz, 1987). For example, the impact of a microsystem influence, such as family emphasis upon academic achievement, may be either strengthened or blocked by nonfamily mesosystem factors, such as the quality of the school system the child attends.

3. To the extent that the environment functions as a system, influences should be *bidirectional* in nature, such that lower levels of the environment should have the potential to influence environmental process at higher levels (Gottlieb, 1991). In addition, given that the environment is differentiated not only across but *within levels,* there also should be bidirectional influences across subunits within a given level.

4. In addition to operating in space, the environment also operates across time (the *chronosystem*). For example, macrosystem characteristics are transmitted across generations by cultural institutions such as family, workplace, or school (Bronfenbrenner, 1989). The environment also operates across time through earlier events influencing or mediating the nature of environment-development relations at later points in time. Discussion of chronosystem influences is presented in Chapters 5 and 8. What is important to keep in mind is that, although the chronosystem operates in a dimension that is different than other aspects of the environmental structure (time rather than space), it still is a part of the structure. Thus chronosystem influences may be mediated by other dimensions of the environmental structure. For example, Crockenberg (1987) has shown that mothers who experienced rejection during their childhood, and who receive little current support, are more likely to be characterized as angry or punitive parents than mothers who experienced rejection during childhood but who have high levels of current social support.

The above characteristics tell us how the structure of the environment *should* function to influence children's development. Given that we live in an imperfect universe (Dr. Pangloss notwithstanding), it becomes legitimate to ask: Does the structure of the environment function in the ways proposed? Evidence on this question is presented in the next sections.

Hierarchical Environmental Influences

CAREGIVER VALUES AND BELIEFS

The macrosystem level. If environmental systems models are valid, we would expect to see parallels between macrosystem values and goals and caregiver values and goals. Goodnow (1988) has reviewed much of the cross-cultural evidence and has concluded that there is greater variability in caregiver values and goals across cultures than there is variability within cultures. For example, Caudill and Weinstein (1969) note that in Japan the infant is viewed as a separate organism who must be taught to become part of the group, whereas in the United States the infant is seen as a dependent organism who needs to be taught independence. Similarly Goodnow, Cashmore, Cotton, and Knight (1984) compared parental beliefs in three cultural groups concerning the ages at which certain skills should occur. Congruent with a macrosystem emphasis on emotional control, Japanese mothers expected earlier signs of social maturity; congruent with a macrosystem emphasis on the importance of social relations and independence, both Australian and U.S. mothers expected verbal maturity and social skills to appear at an earlier age. Congruent with a cultural emphasis on fatalism, Lebanese mothers reported that children's skills will be acquired when they are needed. Goodnow interprets these findings to mean that parent values and goals are "received knowledge," deriving from the macrosystem.

The impact of macrosystem upon parent values and goals also occurs for smaller macrosystems within a given culture. For example, Fry (1988) has compared parent values and goals in two communities in Southern Mexico, one of which was quite peaceful and one of which was characterized by violence. Parents in the more aggressive community assumed that a certain amount of fighting was in the nature of children and that parents could have little influence in this area of development. In contrast, in the more peaceful community, parents felt that they had an influence over their children's level of fighting and that it was the parents' role to control children's aggression.

While the available evidence generally indicates parallel relations between macrosystem and microsystem values, there are some exceptions to this rule. For example, contrary to what might be predicted, Lin and Fu (1990) note that native Chinese families had stronger beliefs about the need to encourage independence for their children than did U.S. parents or Chinese American parents. Possible reasons for these types of discrepancies will be noted later.

The exosystem level. It has been hypothesized that workplace values and demands will influence the parent's own value systems (Belsky, 1984). Supporting this hypothesized linkage, available evidence suggests that entrepreneurial work situations are associated with parents who value achievement and striving in their children, whereas parents who work in bureaucratic jobs tend to value interpersonal skills in their children. Similarly, work situations that allow employees to have some control over their own work activities predispose to a democratic or authoritative child rearing style (Crouter & McHale, in press). While these results suggest parallels between parents' work situations and their own child rearing values, available studies have not clearly distinguished whether work values shape parent values or whether parents with certain values gravitate into certain types of occupations.

CAREGIVER BEHAVIORS

The macrosystem. There are a number of examples of clear parallels between parent behaviors and macrosystem values and goals. For example, Dixon, LeVine, Richman, and Brazelton (1984) report clear parallels between macrosystem beliefs emphasizing obedience and meeting societal expectations, and the degree of

maternal praise or modeling of tasks when African mothers interact with their infants in teaching situations. Similarly, noting that
object exploration is valued less than the exchange of objects
among Kung bushmen, Bakeman, Adams, Konner, and Barr (1990)
report that Kung adults appear to discourage infant object play that
does not involve giving. For older children, Fry (1988) reports that
Mexican parents in a more aggressive community do not discourage children's fighting, whereas in a more peaceful community
both children's actual fighting and play fighting were discouraged
by parents.

Parental behaviors may be particularly sensitive to macrosystem-
driven beliefs about when infants or young children should develop certain skills and what parents can do to promote the development of these specific skills. For example, Kindermann and
Skinner (1988) have argued that cross-cultural differences in beliefs about the importance of early motor development are related
to the ages at which motor behaviors are taught to children.
Similarly LeVine, Miller, and Richman (1991) report a significant
correlation between Mexican mothers' estimates of the age at
which they felt infants could recognize their mother's voice and
the mother's verbal responsiveness to her own infant.

Taking the cross-level process one step further, Rogoff (1990)
emphasizes the importance of *guided participation* (also called *apprenticeship* or *scaffolding*) as the means through which children
learn macrosystem-appropriate beliefs, goals, and problem-solving
mechanisms. Guided participation occurs when adults or older children build "bridges from children's present understanding and
skills to reach new understanding and skills" through "arranging
and structuring children's participation in activities" (Rogoff, 1990,
p. 8). The form of guided participation may depend upon the
nature of the macrosystem. For example, in macrosystems where
children do not participate in adult work until late adolescence,
parents may guide their child by providing explanations or by
labeling concepts. In macrosystems where even young children
contribute to adult work situations, parents may be more likely to
use demonstration, reduce tasks down to specific subcomponents

that the young child can handle, or place children in situations where they can observe the ongoing activities of adults as they go about their daily work (Rogoff, 1990). While most obvious for older children, guided participation also may occur in infancy. For example, Rogoff (1990; Rogoff et al., in press) has noted that, in macrosystems that emphasize children adapting to adults, the caregiver may not simplify his or her speech to young children or may respond to children only when the caregiver wishes the child to respond to him or her.

While the above evidence suggests that macrosystem values and goals are translated into differences in parent behaviors, other evidence indicates either inconsistent or even counterintuitive findings. For example, Bakeman et al. (1990) report that, when Kung infants offer objects to adults (which fits cultural values), adults do not show obvious encouragement of these culturally valued activities. Similarly, comparing parent behavior in Japanese and U.S. families with developmentally disabled children, Nihira, Tomiyasu, and Oshio (1987) report that, while some expected differences occur (e.g., U.S. families scored higher on emphasizing warmth and independence), nonexpected differences also occurred (e.g., Japanese families scored lower on academic stimulation, control, and emphasis on family harmony).

Further, while Rogoff (1990) has provided a variety of obvious examples of guided participation, many nonobvious activities also are labeled as guided participation, such as children's solitary activities. Some studies also question the universality of guided participation. For example, Turkheimer, Bakeman, and Adamson (1989) note that guided participation in infancy may occur only when mothers are specifically instructed to interact with their infants. Similarly, Bakeman et al. (1990) note that scaffolding theorists argue that joint attention to an object with a nurturing language user is a critical condition for the facilitation of language acquisition. In the Kung bushman setting, however, infants rarely receive this type of interaction, yet Kung infants show a high level of language use. Potential causes for these types of inconsistencies will be discussed later in this chapter.

The exosystem. It has been argued that exosystem stressors may directly influence caregiver behavior through interfering with the caregivers' ability to deal effectively with their children (Bell & Chapman, 1986). Available evidence does suggest a negative relation between the amount of stress encountered by the family and the quality of parenting (Crnic, Greenberg, Ragozin, Robinson, & Bashman, 1983; Crouter & McHale, in press; Wahler & Dumas, 1989). The impact of family stress may influence not only the microsystem but also the mesosystem given that, in families reporting more stress in their lives, children tended to be in lower quality day care (Howes & Stewart, 1987). What is not clear is whether there are direct or indirect relations between exosystem stress and the environment of the child at the meso- and microsystem levels. In reviewing the available evidence Crouter and McHale (in press) note that relations between work stress and measures of the quality of parenting appear to be mediated by a variety of factors, such as which parent feels stress on the job, the age of the child, and the educational level of the parent. Work stress also may be mediated by whether parents are satisfied or dissatisfied with their jobs (Belsky, 1984).

The mesosystem. Direct transmission from mesosystem to microsystem is most often viewed as occurring through the mechanism of *social support*: the social network of the caregiver (friends, spouses, or neighbors), which provides suggestions, information, or emotional support. It has been argued that parents with larger support networks may get more advice from friends or neighbors as to how to best deal with their children or may receive assistance along such dimensions as extra monetary support or baby-sitting, which may help relieve parental stress (Belsky, 1984).

While a number of studies report a positive relation between level or type of social support and the amount of affectionate sensitive parenting that is observed (Belsky, 1990; Cotterell, 1986; Crnic, et al., 1983; Volling & Belsky, 1991), available results are not completely consistent. Both Crnic and Greenberg (1990) and Crockenberg (1987) report no relation between maternal report of support and the quality of caregiving with young children. One possible explanation for these discrepant results is the possibility that environmental support acts as a "buffer" (Rutter, 1983b), protecting the individual against stress *only when stress is present.* Data by Crockenberg (1981), indicating that maternal perception of support relates to infant attachment primarily for highly irritable infants, is congruent with this hypothesis, as is more recent data by Crnic reporting interactions between stress and support (Crnic & Greenberg, 1990; Crnic et al., 1983).

While social support may relate to parent behavior, particularly during times of stress, some caution must be exercised in interpreting this relation. Caregivers who are capable of providing sensitive, affectionate child care also may be more skilled in relating to other individuals in the community and may therefore have a larger social network (Belsky, 1990). Thus, rather than social support influencing parenting, individuals who are high quality parents may be more likely to have larger social networks. Supporting this argument is evidence by Crockenberg (1987), indicating that mothers who were rejected during their own childhood were much less likely to have supportive relations with other adults. Thus the current pattern of results suggests a more complex pattern of relations between the microsystem and the mesosystem than may be encompassed by a simple support-behavior relation.

SUMMARY: HIERARCHICAL INFLUENCES OF THE ENVIRONMENT

In general, available evidence indicates that the environment at higher levels can directly and indirectly influence the nature of the environment at lower levels. There also have been a number of exceptions to this general pattern, however, particularly when going from the macrosystem to the microsystem. One possible explanation for these divergent results is the possibility that macrosystems are not static and that there may be changes across time in central macrosystem beliefs, values, or goals. At least for traditional cultures one major influence causing change is the degree of urbanization or Westernization in the culture. Urbanization or Westernization could lead to within-macrosystem divergences, depending upon whether individuals in the macrosystem followed traditional or modern beliefs about rearing (Goodnow, 1988; Schaefer & Edgerton, 1985). For example, Ho and Kang (1984) report that there are few differences in child rearing attitudes expressed by Hong Kong mothers and grandmothers. There are a number of differences between grandfathers and fathers, however, with grandfathers showing a higher degree of traditionalism in their child rearing attitudes. Lin and Fu (1990) have noted that, while Chinese American mothers emphasize control of the child more than do American mothers, they emphasize control less than Chinese mothers living in Taiwan. LeVine et al. (1991) report that the higher the level of schooling of Mexican mothers, the earlier they believe their infants can recognize the mother's voice. These group differences suggest a gradual change in parental attitudes and behaviors due to acculturation.

An alternative hypothesis is based on the possibility that the macrosystem may not be homogeneous. Rather the macrosystem may have different subunits, similar to those found at other levels. Thus caregivers may be simultaneously living in several macrosystems, each of which may influence the way in which individuals respond to specific macrosystem beliefs. One example of this process is noted by Goodnow (1988), who reported that Israeli Kurdish mothers showed a good grasp of modern medical principles, in part because traditional Kurdish views of illness were in physical terms. In contrast Israeli Yemeni mothers had difficulties grasping modern medical principles because their traditional explanations of illness were primarily magical in nature. Thus the degree of "fit" between the individual and the variety of macrosystems he or she lives in may influence response to individual macrosystem pressures.

Additionally, within a systems perspective, it must be recognized that the macrosystem does not exist in isolation. While the macrosystem no doubt influences the microenvironment, influences from intermediate levels also may influence the micro environment. For example, Nihira, Mink, and Shapiro (1991) reported that, in addition to cultural differences, more Asian-American mothers were employed than Euro-American mothers, and the Asian-American mothers worked longer hours for lower pay. In addition, in Asian-American families the child care network primarily involved extended family, whereas in Euro-American families the child care network involved friends or nonfamily members. Thus the impact of traditional macrosystem values and beliefs upon the microsystem may be attenuated or strengthened by intermediate-level environmental influences, such as family support or work stress.

Along the same lines available evidence suggests that it may be important to consider meso- and exosystem factors in combination rather than in isolation. For example, looking at the impact upon maternal behavior of father's absence as a function of employment, Cotterell (1986) has presented results suggesting a chain of influence. Specifically the father's work pattern, while not predicting maternal child rearing patterns, does influence the mother's social support network (smaller and more limited networks for those women whose husbands are absent due to employment conditions). Support networks in turn influence the quality of the child's rearing environment. Similarly the influence of quality of marital relationship upon degree of paternal involvement in child caregiving activities appears to differ in dual- and single-wage-earner families (Crouter, Perry-Jenkins, Huston, & McHale, 1987). An interaction between stress and support is seen in the data by Billings and Moos (1983), indicating a 15% drop in the incidence of children's disturbance in families where there are depressed parents but where there is also a combination of low stress and high support.

The existence of multiple interacting paths of influence suggests that our ability to understand how the structure of the environment works will be enhanced when we go beyond looking at individual linkages to consider systemwide links. The next critical question concerns whether the complex interconnections between different levels and subsystems of the environment have implications for understanding the nature of environmental influences at a specific level of the environment. That is, can environment-development relations at one level be mediated by the interconnections between different levels of the environment? The answer to this Phase III question is found in the next section.

Mediation of
Environment-Development Relations

MEDIATION FROM THE MACROSYSTEM

The possibility that macroenvironmental parameters may mediate between the microenvironment and development has been noted in a number of recent reviews (Bronfenbrenner, in press; Wachs, 1991a). One means through which this mediation might occur is when the individual's perception of stimuli, or the meaning of stimuli at the microenvironmental level, changes in different macrosystem contexts (Bronfenbrenner, in press). For example, Caudill and Weinstein (1969) suggest that maternal verbal behavior may have different functions in Japanese and American culture, with verbal behavior functioning to soothe the baby in Japan and stimulate the baby in the United States. Similarly Asian children appear to view high levels of parental control as an expression of parental love, whereas the same meaning is not given to parental control in other populations (Bronfenbrenner, in press).

What little evidence is available does support the hypothesis that macrosystem context may act as mediator. For the most part this evidence is based on research indicating that microenvironment/outcome relations that appear for one group of individuals do not appear with individuals from different contexts. Thus Bradley et al. (1989) report that, while significant relations were found between specific microenvironmental parameters and cognitive outcomes for both whites and blacks, there were few consistent relations for Mexican American children, even when the groups were matched for social class. Similarly, whereas significant relations between environment and outcome were found for middle-class children, the magnitude of reported relations was substantially smaller for children in lower socioeconomic groups. Bradley et al. (1989) suggest that these group differences may reflect macroenvironmental-driven differences in the timing of specific parental practices. Similarly, Dornbush, Ritter, Leiderman, Roberts, and Fraleigh (1987) report that there were significant relations between school grades and adolescents' perceptions of their parents' rearing styles (authoritarian, permissive, or authoritative) for both blacks and whites; for Asian American adolescents the correlation of grades with perception of parental styles was basically zero order. Looking cross-culturally, Nihira et al. (1987) reported that parents' acceptance of the child or family emotional expressiveness were significant predictors of emotional adjustment for U.S. children but were generally unrelated to behavioral outcomes for Japanese children.

In addition to the social microenvironment, the impact of the physical microenvironment upon development also may be mediated by aspects of the macroenvironment. For example, Draper (1973) has reported that, even though Kung bushmen live in extremely crowded microenvironments (densities of up to 30 people per room not being uncommon), there was no evidence of stress-related diseases in the Kung population; in contrast, crowding is associated with stress disorders in North America populations (Milgram, 1970). Draper suggests that both the physical ecology (camps are crowded but there is tremendous space between camps) as well as family social support systems act to buffer the Kung against the impact of crowding. Along the same lines Stevenson and Lee (1990) suggest that cultural values stressing the importance of effort and the importance of education may buffer children in Taiwan against the detrimental impact of extremely crowded classroom situations. A similar argument has been made by Lieh-Mak (1980), who suggests that crowding may have different effects on children's development, depending upon the degree of cultural support for high density living.

It also may be important to consider the potential mediating influence of the macrosystem upon mesoenvironmental processes. One example is seen in data noted by Pickles, Rutter, and Quinton (1989) suggesting that military service may have beneficial effects for youths from disadvantaged backgrounds that will not occur for those from nondisadvantaged backgrounds.

While a number of studies indicate the operation of macrosystem mediation, not all studies show this pattern. Similar patterns of relations between caregiver stimulation or expectations and developmental outcomes have been reported for Western and Oriental cultures (Nihara et al., 1987; Stevenson & Lee, 1990). Analysis of the relation of caregiver behavior to toddlers' cognitive and behavioral competence has indicated similar patterns of relations in the United States, Kenya, and Egypt (Wachs et al., 1992). In the United States Luster and Dubow (1991) have noted similar relations between dimensions of the home environment and children's verbal intelligence in Caucasian, black, and Hispanic populations. One possible explanation for these divergent findings has been noted by Steinberg, Mount, Lanborn, and Dornbusch (1991), who suggest that, while there does appear to be evidence for macroenvironmental mediation, in many cases the effect size (strength) of macroenvironmental mediation is low. This could mean that existing mediating effects may be missed in some studies due to low statistical power. An alternative explanation may involve the changing, heterogenous nature of the macrosystem, as noted earlier.

THE MESOSYSTEM

Evidence also is available suggesting the operation of meso-system mediators. Howes (1990) has reported that, for children who were not in child care, family socialization practices had a significant influence on the child's adjustment; for families whose children were enrolled in day care, family socialization practices showed very little prediction of children's adjustment. Similarly Luster and Dubow (1991) related measures of children's home environment to children's verbal intelligence prior to and after the time children entered school. Results indicated consistent relations between home environment and children's verbal intelligence in the preschool children, whereas for school age children the predictive value of home environment on verbal intelligence was clearly diminished.

CONCLUSIONS

Available evidence suggests that there may be mediation by higher levels of the environment of environment-development relations at lower levels. Though these mediating effects may be small, they do exist. Particularly at the microsystem level, relations between specific environmental dimensions and development may be accentuated or diminished, depending upon processes occurring at higher levels of the environment.

Bidirectional Influences

Some of the strongest evidence for within-level bidirectionality has come from the study of a relatively neglected dimension of the microsystem, namely, the physical microenvironment. While the overwhelming majority of research on microsystem influences has involved the social microenvironment—that aspect of the environment encompassing the nature and extent of relations between children and caregivers—at least one other major aspect of the microsystem exists, namely, the physical environment (Wachs & Gruen, 1982). There have been a number of studies demonstrating relations between various aspects of children's development and specific dimensions of the physical microenvironment (see Chapter 3; for more detailed reviews of physical environment-development relations, see Evans, Kliewer, & Martin, in press; Wachs, 1986; Weinstein & David, 1987; Wohlwill & Heft, 1987). At its simplest level the physical microenvironment may be defined as the stage or setting upon which occur caregiver child relations. More detailed taxonomies of the nature of the physical microenvironment have been offered by both environmental psychologists (Canter, 1977; Evans et al., in press; Moore, 1987) and developmental psychologists (Wachs, 1989; Wohlwill & Heft, 1987). For the most part three major dimensions seem to characterize available taxonomies. These include (a) *spatial characteristics*, such as open versus closed space or crowding (David & Weinstein, 1987); (b) *"affordance"* characteristics of the environment, such as objects that afford the child specific uses (Wachs, 1986); and (c) *"affordanceless"* aspects of the environment, which typically involve nonspecific background stimulation such as noise (Wachs, 1986).

While caregivers clearly can influence the nature of the child's physical microenvironment, as in controlling noise levels or buying age-appropriate toys, available evidence also suggests a *bidirectional process*, with certain dimensions of the physical microenvironment influencing the nature of caregiver-child relations. Data from cross-cultural research (Kaplan & Dove, 1987; McSwain, 1981) have clearly indicated that physical ecological characteristics may mediate the nature of relations between caregivers and children. For example, hazards in the natural environment may influence the degree to which parents allow children to independently practice such skills as walking or exploration (Kindermann & Skinner, 1988; Moos, 1973). Available evidence from developmental research also suggests that specific dimensions of the physical microenvironment may act to inhibit the expression of a variety of critical caregiver behaviors, such as sensitivity, responsivity, language stimulation, and guided participation. The critical physical dimensions that have been identified as inhibitors involve *noise* and *crowding*. In a series of studies using infants ranging in age from 3 through 24 months of age, we have consistently demonstrated that infants and toddlers being reared in noisy, crowded homes have caregivers who are coded as being less responsive, less involved, less vocally stimulating, less likely to show or demonstrate objects, and more likely to interfere with toddler exploratory activities (Wachs, 1986, 1989; Wachs & Camli, 1991; Wachs & Desai, 1992). Similar patterns of relations also have been reported by other researchers (Gottfried & Gottfried, 1984; Hannan & Luster, 1990; Ricciuti & Thomas, 1990; Shapiro, 1974; Woodson & DaCosta-Woodson, 1984).

It could be argued that the above results may be artifactual, perhaps being due to exo- or mesosystem factors such as work stress or availability of support. In a direct test of this hypothesis, however, Wachs and Camli (1991) reported that predicted relations between noise or crowding and parental behavior were basically unaffected when meso- and exosystem factors such as stress and support were partialed out of this relation. The fact that caregivers can directly influence the physical setting within which their children function, while the physical setting in turn can influence how caregivers behave toward their children, nicely illustrates the bidirectional nature of the environmental system at a given level of the environment.

BIDIRECTIONALITY ACROSS LEVELS

Is there evidence that lower level environmental processes can influence higher level environment-development relations? At least one study suggests the possibility of processes occurring at the exosystem level mediating the influence of macrosystem factors. Using a sample of Lebanese children living in Beirut, Day and Ghandour (1984) looked at the impact upon children's aggressive play behavior of watching violent versus nonviolent television. Given that Middle Eastern values support the expression of aggression by boys and inhibit the expression of aggression by girls, it is not surprising that Day and Ghandour report more aggressive behavior by boys following viewing of aggressive TV programs. The *one exception* to this pattern was a group of children for whom another round of citywide fighting had broken out, just prior to the time these children were due to be tested. Both boys and girls in this group become more aggressive following exposure to TV violence, and there were *no differences* between the sexes. The authors hypothesize that the impact of citywide sectarian violence may have interfered with the sex-linked relation between cultural values and aggression, suggesting an influence of the exosystem upon macrosystem processes.

Some evidence also is available suggesting that microsystem processes can moderate both exosystem and mesosystem influences. Investigating the relation of maternal employment patterns to children's school competence, Moorehouse (1991) reported that children whose mothers went from part- to full-time employment had lower cognitive and social adjustment than did children whose mothers were home full time or children whose mothers continued to work full time. For children whose mothers had shifted from part- to full-time employment, and who had prior history of frequent shared activities with their children, however, the shift from part- to full-time employment did not result in decrements in their children's cognitive or social competence. These results suggest the mediating impact of a microsystem process (shared mother and child activity) upon the impact of an exosystem dimension (job characteristics). Evidence for a reverse path going from the microsystem to the meso system is seen in the data by MacDonald and Parke (1984), who reported a link between the quality of maternal and paternal interactions with children and the children's subsequent interaction with peers in school. MacDonald and Parke suggest that appropriate interaction with family members may provide the child opportunities to learn and practice social skills, which in turn will influence how the child functions in the school environment.

The available evidence, while scarce, has demonstrated reverse linkages between the exo and macrosystems and between the micro and exo and mesosystems. I have been unable to find evidence indicating that microsystem processes may influence the macrosystem. The possibility has been raised, however, that microsystem factors may influence the macrosystem across time. Noting that some behavioral differences between infants in different cultural groups appear at birth, several researchers have suggested that the infants' predisposition to respond in certain ways may shape caregiver behaviors toward them, which ultimately may influence cultural values about appropriate parental practices (Freedman, 1974; Lester & Brazelton, 1982; Shand & Kosawa, 1985). While these early differences in infant behavior could suggest the possibility of a link going from microsystem to macrosystem, there are alternative explanations. Specifically Chisholm (1981, 1983) has presented evidence indicating that these early neonatal differences may be due to differences in the fetal environment, which in turn may be a function of ecological or macrosystem factors such as maternal diet, age of the mother when giving birth, and drugs used during pregnancy. Thus the possibility of a reverse link from microsystem to macrosystem, while intriguing, must still be considered speculative at this point.

Summary: The Structure of the Environment

If nothing else, it should be clear that the structure of the environment is a complex, multifaceted system. While it is important to have detailed study of specific elements of this system, ultimately we will need to simultaneously investigate multiple levels of the environment to understand how the environment operates. The focus of future Phase III investigations should involve understanding how the different levels of the environment might influence each other, influence development, and mediate environment-development relations at the various levels.

Note

1. While the macrosystem is the highest level in Bronfenbrenner's model, it is possible to go beyond the macrosystem in the sense that ecological factors such as landscape features or climate may influence macrosystem beliefs.

5

Operation of the
Environment Across Time

In Chapter 4 I discussed how the structure of the environment can be viewed as having a temporal component. In this chapter the focus is on both time-limited and longitudinal aspects of environmental action.

Critical and Sensitive Periods

Historically one of the earliest lines of investigation on temporal aspects of environmental action has to do with the concept of the "critical period" and the phenomenon of "imprinting" (Colombo, 1982; Scott, 1979). The traditional view of critical periods, deriving from embryology, is that during certain periods of development the organism is unusually sensitive to specific types of stimulation. Specific types of stimulation, if encountered during this critical time frame, will have long-term influences upon development that are not likely to be modified by later stimulation. In contrast the same stimulation occurring before or after this critical time frame is not likely to have a major impact upon development.

For example, even a brief exposure to a member of one's species during the critical period for socialization is said to channel the organism's social preferences for the remainder of its existence (Scott, 1979). Evidence for the existence of critical periods has been well documented in infrahuman species (Bateson, 1979; Bornstein, 1989b; Colombo, 1982), though some reviewers have argued that critical periods are the exception rather than the rule in mammals (Lamb & Hwang, 1982) or that critical periods may not be as sharply defined a phenomenon as is generally assumed (Bateson & Hinde, 1987). Claims also have been made for the existence of human critical periods in such domains as language (Scott, 1979), personality (Beit-Hallahmi, 1987), and social relations (Moore & Sheik, 1971). For example, it has been argued that the first few hours after birth are a critical period for the formation of mother-infant bonding and that, if mother and infant are separated during this time, the bonding process may be seriously disrupted; while this hypothesis has received wide publicity, available evidence does not support the existence of a critical period for mother-infant bonding (Myers, 1987). Overall, available evidence generally indicates that there is little that is analogous to the traditional critical period phenomena for human behavioral development (Bornstein, 1989b; Colombo, 1982; Lamb & Hwang, 1982; Rutter, 1985b).

THE CONCEPT OF SENSITIVE PERIODS

Instead of critical periods researchers have preferred to use the concept of "sensitive" or "optimal" periods when referring to human development. These terms refer to a time frame when the individual may be more or less sensitive to specific aspects of the environment than at other times (e.g., Benjamin, 1965; McCall, 1981). In contrast to critical periods sensitive periods are relativistic rather than deterministic; that is, the length of time defining the sensitive period is more flexible; more than a brief exposure to specific stimulation is necessary; only certain developmental outcomes may be influenced; and the long-term impact of exposure during sensitive periods may be attenuated or strengthened by later environmental influences (Bornstein, 1989b; Oyama, 1979). What is common to both concepts is the question of *when* the organism is more or less sensitive to the environment. Viewed in this way both critical and sensitive period studies can be seen as examples of Phase II research, with a focus on when rather than what.

Does available evidence support the existence of sensitive periods for human behavioral development? Almli and Finger (1987) have noted that the central nervous system may be maximally sensitive to diffuse neural damage or undernutrition during the period of most rapid brain development. Aslin (1981) has provided evidence suggesting the possibility of early sensitive periods for the development of human binocular vision and certain auditory sensitivities. Other domains in which potential sensitive periods may exist include learning sign language (MacDonald, 1986) and the development of speech perception (Snow, 1987). Environmental inputs that also may be influenced by the operation of sensitive periods include developmental intervention with handicapped children (Casto, 1987) and vulnerability to separation experiences such as hospitalization (Rutter, 1981a).

Other evidence is less supportive of the concept of sensitive periods. For example, Epstein (1976) has argued that central nervous system growth spurts result in sensitive periods for learning and cognitive development. McCall, Myers, Hartman, and Roche (1983), however, have presented evidence that does not support the validity of Epstein's hypothesis of specific sensitive periods for learning or cognitive development. Grych and Fincham (1990), in a review of the relations of marital conflict to children's adjustment, conclude that, while children at different ages use different coping mechanisms and are differentially aware of the existence of marital conflict, no single age group appears to be especially vulnerable to the impact of marital conflict. Almli and Finger (1987) have presented evidence indicating that there does not appear to be a sensitive period for recovery from specific localized central nervous system damage, while Snow (1987) has clearly documented that, with the exception of speech perception, there does not appear to be a sensitive period for language development.

Even if sensitive periods do occur for humans, there appears to be little agreement as to when these periods might be. MacDonald

(1985) has argued that as children get older their development may become less open to environmental influences, which would suggest that sensitive periods are more likely to occur early in the life span. In contrast McCall (1981), in his "scoop" model, has argued for a high level of genetic control over development across the first several years of life, with the infant following a highly canalized species-specific developmental path that minimizes any long-lasting environmental influences. This would suggest a later rather than an earlier sensitive period. While some evidence is supportive of the scoop model (e.g., Bakeman & Brown, 1980), a number of studies report significant relations within the first six months of life between specific aspects of the environment and early cognitive development (Bornstein & Tamis-LeMonda, 1990; Yarrow et al., 1975).

While an appealing idea, the available evidence does not suggest the validity of the sensitive period concept as an *overall model* for explaining environmental contributions to human development (Bornstein, 1989b). In part the pattern of conflicting results in this area may be due to differences between studies in the quality of the methodology used to study sensitive periods. Bornstein (1989b) has defined 14 parameters that must be specified to adequately test for the operation of sensitive periods (e.g., defining when and how often the proposed sensitive period occurs, how long it occurs, what are the nature and origins of the relevant experiences during the sensitive period, and what are the nature and duration of the outcome of exposure to specific stimuli during a potential sensitive period). Particularly at the human level, few if any studies contain the detailed information that Bornstein sees as necessary for a satisfactory test of sensitive periods. This is particularly true for social address or epidemiological studies. For example, looking at mating patterns of individuals who were kibbutz reared, Shepher (1971) suggested that there was a sensitive period during the first six years of life for learning with whom not to mate. Analyzing these results Bateson and Hinde (1987) point out that age of entry into the kibbutz is clearly confounded with length of exposure to potential mates, making it impossible to determine whether the observed results were a function of the operation of a sensitive period or just due to cumulative experiences. Intervention studies, wherein interventions begin at different age periods, might seem an obvious approach to this question. As MacDonald (1986) has pointed out, however, the age at which intervention is started often is confounded with length of intervention, again making it difficult to ascribe results solely to the operation of sensitive period phenomena.

Conceptually the traditional view of sensitive periods also may be overgeneralized when applied to human development. Rather than overall sensitivity, evidence suggests that sensitive periods may be important only for a limited set of specific developmental outcomes (MacDonald, 1985). For example, Oyama (1979) has argued that very early maturing peripheral systems are more likely to be period sensitive than are more complex, later maturing systems. The critical question is why some developmental outcomes appear to be influenced by sensitive periods whereas others are not.

One potential answer to this question comes from the field of developmental neuropsychology. Based on studies of environmental contributions to central nervous system development, Greenough (Greenough, Black, & Wallace, 1987) offers the intriguing suggestion that there may be two classes of developmental outcomes: *experience expectant development* and *experience dependent development*. Neurologically the operation of experience expectant development is seen in the central nervous system generating an excess number of synaptic connections early in life; the operation of experience dependent development is seen in the pruning back and localization of excess synaptic connections by later environmental encounters (Greenough et al., 1987). Experience expectant development is evolution based, in the sense that there is a relative advantage if individuals are "wired" to take advantage of experiences that have been traditionally encountered by members of their species at certain developmental periods. Some of these species-typical experiences may occur through the behavior of the child. For example, human infants may be "prewired" to become attached, given that infant distress cries have reliably elicited adult attention across the evolutionary history of our species. In contrast experience dependent development involves the storage of specific information that is unique for each individual. Within this framework those aspects of development that are experience expectant would be more likely to be influenced by sensitive periods, whereas those aspects of development that are experience dependent would be less likely to involve sensitive periods (Greenough et al., 1987).

Although Greenough's analysis is based primarily upon infrahuman data, there are parallels at the human level. Oyama (1979) has suggested that human sensitive periods may occur only for circumstances that have a high probability of occurrence. Similarly, Bornstein (1989b) has hypothesized that human sensitive periods may be adaptive for individuals who are likely to continually encounter species-typical environments, but would be maladaptive for individuals likely to encounter species-atypical environments. Thus a phenomenon such as attachment or language may be more experience expectant, given that most young children are likely to encounter responsive language-using adults very early in life. In contrast specific types of cognitive or social skills may be more experience dependent, given the varied ecological or cultural conditions under which children develop and function. Thus, rather than looking for general sensitive periods, it may be more appropriate to restrict the sensitive period model only to those aspects of human development that are more likely to be experience expectant. By integrating the concepts of sensitive periods and experience expectant stimulation, we have moved beyond a Phase II descriptive approach to a Phase III process question about the nature of environmental influences at specific time periods.

Going Beyond the Sensitive Period: Age Specificity

Rather than hypothesizing an overall greater or lesser sensitivity to the environment, the age specificity hypothesis assumes a multidimensional environment and predicts that different aspects of the environment will become relevant for development at different ages. Age specificity differs from sensitive period approaches in that no one developmental period (or periods) is viewed as being more or less sensitive to environmental influences. Rather, within the age specificity framework, what differentiates age periods is not degree of sensitivity but the types of stimulation to which the organism is sensitive. While integrating what and when, age specificity basically is a Phase II approach to the study of environmental influences.

The genesis of the concept of age specificity is seen in the careful observations made by psychoanalytic clinicians of differential reactivity by young infants to different forms of stimulation at different ages (Escalona, 1968; Spitz, 1965). Previous reviews have documented a variety of age differences in reactivity to specific stimulation (Wachs & Gruen, 1982; Yarrow & Goodwin, 1965). For example, tactual-kinesthetic stimulation has been implicated as being especially salient for development primarily in the first six months of life, whereas verbal stimulation is seen as having relevance for performance on standard infant tests of cognitive development only after 12 months of age.

More recent research continues to provide evidence for the existence of age-specific environment-development relations. Research continues to show that tactile stimulation is positively related to early cognitive development and negatively related to later cognitive development, whereas caregiver vocal responsivity is unrelated to early cognitive performance and positively related to later cognitive performance (Coates & Lewis, 1984). Bradley and Caldwell (1984) have presented evidence indicating that maternal responsivity may be particularly critical for early rather than later cognitive development. Bornstein (1989c) reviews evidence suggesting that maternal instructional techniques show an increasing impact upon children's cognitive development, while maternal attentiveness and emotionality become less important for cognitive development across the first four years of life.

Evidence for age specificity also exists at levels of the environment other than the microsystem. At the macrosystem level Rogoff (1990) has suggested that, in cultures where peer or sib rearing is common, older children may benefit more than younger children from peer or sib interactions. At the exosystem level

Crouter and McHale (in press) present evidence suggesting that parental absence from the home due to work pressures may be more of a negative influence upon younger children than older children, with older children possibly even benefiting from the increased autonomy associated with having two working parents. At the mesosystem level Crnic et al. (1983) suggest that the direct impact of parental social support networks upon children's development may increase as children get older.

While there is little direct evidence, there are a number of possible reasons why age-specific environment-development relations may occur (a Phase III question). Changes in the central nervous system may influence the kinds of stimuli that children are capable of processing (Fabiani, Sohner, Tait, & Bordieri, 1984; Werner & Lipsitt, 1981). For example, Shaheen (1984) used existing evidence on central nervous system development to successfully predict which developmental functions might be most affected by exposure to lead at different age periods. In addition to changes in central nervous system development, age-associated changes in hormonal patterns also may influence how children react to stimulation at different ages (Brooks-Gunn & Warren, 1989).

Changes in children's functional capacities or processing abilities may be another age-related parameter that mediates differential reactivity to environmental stimulation at different ages. There are a variety of potential functional parameters that could lead to differential environmental influences at different ages. These include age-related differences in *selective attention* (Ackerman, 1986; Enns & Akhter, 1989; Hamilton, 1983; Lane & Pearson, 1982), *stimulus thresholds* (Berg, 1975; Berg & Smith, 1983), *discriminative abilities* (E. Cummings, Vogel, Cummings, & El-Sheikh, 1989; Irwin, Stillman, & Schade, 1986), the ability to *regulate emotional expressiveness* (Hyson, 1983) or *ongoing behavior* (Reed, Pien, & Rothbart, 1984), the emergence of the ability to *anticipate consequences and rewards* (Bronson, 1985; Gunnar, 1978), and the ability to use *information from multiple sources* (Millar, 1984; Sophian, 1986).

While there is evidence supporting the existence of age-specific environment-development relations, and a number of reasons why such age specificity should exist, certain cautions are warranted. In many cases it is difficult to distinguish between random results and genuine age-related differences in environment-development relations. The major reason is that many studies do not predict ahead of time what environment-development relations would be expected at a given age. Without such specific a priori predictions, chance age differences could be used as evidence for age specificity. For example, Abraham, Kuehl, and Christopherson (1983) report that parental control strategies are negatively related to the development of empathy for 3-year-old children, whereas paternal control strategies are positively related to empathy for 5-year-olds. Given the lack of a priori predictions, it is not clear whether these complex findings reflect genuine age specificity or random age differences.

One obvious way of getting around this problem is to replicate previous results showing age-related environmental influences. Certainly there appears to be consistent evidence indicating that physical contact stimulation is primarily related to development early in life, whereas verbal stimulation shows an increasing influence upon development as the child grows older (Bornstein, 1989c). A second way of approaching this problem is to use what we know about age-related changes in children's functional competencies to predict what aspects of the environment should be related to development at different ages. One excellent illustration of how this might be done is seen in a paper by Wasserman (1984), wherein he attempts to predict what types of cognitive behavior therapy might be most appropriate for children at different ages. Based on prior knowledge of children's cognitive development, Wasserman hypothesized that, for young children who do not have adequate language or verbal representational skills, modeling should be the most appropriate therapy technique. When children develop internalized speech patterns, self-instruction techniques may be particularly appropriate. For adolescents who have reached the formal operational stage, the preferred choice would be therapeutic techniques, such as rational emotive therapy, which require the child to process verbal information and test hypotheses. Another potentially testable example is seen in evidence indicating that young children may be particularly handicapped in detecting adult speech in a noisy environment because of their lowered ability to filter out language signals from background noise (Irwin et al., 1986). This would suggest that the detrimental impact of environmental noise upon language development may be lessened as children get older.

Even if such a priori predictions are made, it will also be critical to distinguish between genuine age differences in reactivity to environmental stimulation versus differences in the type or amount of stimulation provided to children at different ages. It is clear that caregivers will systematically shift their behavior toward children as a function of age-related child behaviors, such as the child's level of competence (Heckhausen, 1987a) or the child's negotiation skills (Kuczynski, Kochanska, Radke-Yarrow, & Girnius-Brown, 1987). Thus, in addition to asking whether there are real differences in children's reactions to stimulation at different ages, it is equally critical to assess whether age-related changes in children's functional capacities influence the nature of the children's environments.

Early Versus Later Versus Cumulative Experience

In terms of temporal aspects of environmental action, the question that has probably generated the most controversy is whether early experiences have a long-term impact upon development. Three seemingly mutually exclusive models have been proposed:

Model 1. Sparked by evidence from such diverse disciplines as ethology, psychoanalysis, and neuropsychology, many researchers have concluded that early experiences can have long-term effects upon subsequent development (Wachs & Gruen, 1982): "The extent and manner to which an organism reacts to later environmental events is determined to a large extent by environmental conditions which prevail during infancy" (Levine, 1962, p. 46).

Model 2. Influenced by case studies indicating recovery of function from early severe deprivation, or early intervention studies that fail to show long-term effects, other researchers conclude that environmental influences closer in time to when outcome variables are measured can override the impact of earlier experiences (Clarke & Clarke, 1979; Kagan, 1984): "Early psychosocial experience by itself has no necessary long term effect on later development" (Clarke & Clarke, 1989, p. 289).

Model 3. Other researchers have argued that environmental influences can best be viewed as a cumulative phenomenon, in the sense that initial environmental influences that are stabilized across time by a consistent environment can have long-term influences (Bloom, 1964): "The pattern of high relationships and hence stability manifest in successive trait measures may really reflect the constant and cumulative effect of the environment" (Hanson, 1975, p. 479).

Unfortunately tests of these three competing models are all too rare. Many studies that purport to test one of the three models are inappropriate due to poor methodology. For example, studies that show significant correlations between early experience and later development, but that do not measure the intervening environment (e.g., Coates & Lewis, 1984), are unable to tell us whether observed results are due to a unique influence of the early environment or the continuity of the environment across time. Enough evidence is available, however, to suggest that none of the three models is totally invalid.

Supporting the validity of *model 1* (early experiences having unique long-term effects), experimental research has shown that the occurrence of a unique single experience at 6 months of age can influence subsequent memory performance two years later (Perris, Myers, & Clifton, 1990); given the nature of the 6-month experience (an auditory localization task in an experimental laboratory), it is unlikely that the child encountered subsequent experiences of this type after the initial experience. Looking at the functioning of school age children who were either securely or poorly attached in infancy, Sroufe, Egeland, and Kreutzer (1990) report that measures of the quality of the child's home environment at 30 months of age predict elementary school functioning, even when the impact of intervening home and school environment was statistically controlled. Sroufe et al. argue that one reason for the stability of early environmental influences may be that young children living in more adequate environments not only may be more resistant to later stress but also may be more responsive to later positive features of the environment.

Supporting the validity of *model 2* (the primacy of later environment), Casto (1987) presents data suggesting that, at least for handicapped children, later environmental interventions may be more important for development than are early interventions. Similarly Clarke and Clarke (1986) summarize data suggesting that later exposure to stress can adversely influence development as much if not more than early exposure to stress.

Evidence supporting the validity of *model 3* (cumulative environmental influences) is also available. Laboratory studies have suggested that there is stronger retention of early learning when infants are reexposed to portions of the original learning experience (reinstatement; Rovee-Collier, 1984). Evidence supporting the validity of model 3 is also seen in a study by Olson, Bates, and Bayles (1984). Their results indicate that the relation between 6-month environment and 24-month measures of toddler competence are governed by the stability of the environment between 6 and 24 months of age. Going beyond the infancy period, Weisner, Bernstein, Garnier, Rosenthal, and Hamilton (1990) report that, while infant attachment classification at 12 months of age did not predict 12-year cognitive and emotional performance, the stability of the child's family environment did predict. Children from families with stable life-styles (even if these life-styles were counterculture in nature) had the best outcomes, whereas children from families whose environments were unstable across time showed the poorest adjustment.

What is curious about the above pattern of results is that models 1, 2, and 3 seem to be mutually incompatible, and yet evidence is available supporting the validity of each of the three models. While all three models should not be simultaneously correct, in fact they may be. Both methodological and conceptual factors may lead to this curious state of affairs.

Methodologically a number of researchers have noted the possibility that long-term effects of early environment may appear *only* if later environmental situations reflect or elicit earlier learned behavioral styles or vulnerabilities (Sroufe, 1979). For example, Grych and Fincham (1990) conclude that children may use previously learned behavior patterns only when they encounter later stress. Similarly Sroufe et al. (1990) argue that children with a history of loss or environmental deprivation may function adequately later in life, as long as the children are in supportive environments; it is only when the child is faced with environmental loss or disruption that we would see the operation of earlier environmentally driven vulnerabilities. Thus the impact of early experience, although operating across time, may be revealed only under certain environmental situations. In the case of a child who experienced specific early stress but who did not encounter this type of stress later in life, the outcome would seem to fit model 2, in the sense that the later environment would be said to outweigh the effects of the early stress. This conclusion would change, however, if the child was faced with later stress. In this case the results could fit either model 1 or model 3, depending upon whether the later stress was similar or different to that encountered by the child earlier in life.

A conceptual framework under which all of the above models could be correct is seen in one of the few studies that simultaneously tested the validity of all three models (Bradley et al., 1988). Subjects in this project were 10-year-old children whose home environments had been measured at 6 months, 24 months, and 10 years of age. At 10 years of age measures of the child's school achievement and classroom behavior also were collected. Supporting model 1 Bradley et al. report that 6-month variety of stimulation and parental responsivity predicted 10-year classroom behavior, even after statistically controlling for the effects of 10-year home environment. Supporting model 2 Bradley et al.

report that 10-year parental involvement, family participation, and family emotional climate predicted school achievement, even after statistically controlling for 6- and 24-month home environment. Supporting model 3 Bradley et al. report that the relations between 24-month parental involvement and children's task orientation and school adjustment at 10 years were substantially reduced when the stability of the environmental measures between 2 and 10 years of age was statistically controlled. Bradley et al. conclude that all three models received some support but that none was completely adequate. Bradley et al. (1988) suggest that which model is most valid may be a function of the *outcome and environmental variables that are chosen for study.* As with sensitive periods, no single overall model may cover all situations.

Rather than assuming that early environmental variables show stability or instability of effects across time, or that all outcome variables are equally affected by early, later, or cumulative experiences, the research by Bradley et al. emphasizes the necessity of more specific Phase III predictor models, wherein some early experience variables show long-term effects, others do not, and others show long-term effects only under stable environmental conditions. As with our earlier discussion of age specificity, the trick is to be able to predict ahead of time what specific outcome variables are more likely to be influenced by earlier as opposed to later or cumulative experiences, or what environmental parameters are likely to have long-term, short-term, or cumulative effects. Based on the arguments of Sroufe and his colleagues, it may well be that environmental factors that influence children's early affective and emotional development may be more likely to have long-term influences, over and above the impact of intervening experiences. Similarly, following the work of Bradley et al., it may well be that academic achievement variables are more likely to be influenced by contemporaneous rather than earlier environmental influences. In making these kinds of predictions it is critical to keep in mind the possibility that long-term or cumulative environmental influences may appear only under conditions of later stress.

Summary and Conclusions

While the available evidence suggests that there is nothing akin to a traditional critical period in terms of environmental influences upon children's development, there may well be sensitive periods where the child is more or less likely to respond to existing environmental stimulation. This may be particularly true for those aspects of development that are *experience expectant*. In addition there also may be age specificity, in the sense that different aspects of the environment may become more or less salient for development at different time periods. Clearly what this suggests is that the environmental forces that influence children's development may well change across time (Luster & Dubow, 1991).

In addition it seems clear that the Phase II question of whether earlier, later, or cumulative experience is more important for development is highly oversimplified. Rather the critical question appears to be this: Which developmental outcomes are more likely to be influenced by earlier, later, or cumulative experiences? In answering this question it is important to recognize that we are talking in a probabilistic rather than a deterministic sense. Even for those developmental outcomes that have a higher probability of being influenced by early experiences, there is always the possibility that the impact of early experience can be partially modified by later experience, particularly when later experiences occur in combinations rather than in isolation (Pickles et al., 1989). For those developmental outcomes that are more likely to be influenced by later experiences, it is important to keep in mind that the impact of these later experiences may be filtered through the child's expectancies, many of which are influenced by earlier experiences (Sroufe, 1983). For those outcomes that are primarily influenced by the continuity of the environment, the impact of environmental stability may become apparent *only* when later environmental situations produce stresses or stimulus conditions that are relevant to these early adaptations (Sroufe, 1979). Similarly environmental strategies that are maintained across time by the continuity of the environment may be either maladaptive or adaptive for dealing with later stresses, depending upon the nature of the later stress (Compas, 1987).

Thus, while it is a legitimate Phase II research question to ask which developmental outcomes are more likely to be influenced by early, later, or cumulative experiences, it must always be kept in mind that we are dealing with a Phase III system of influences and not with a set of isolated environmental occurrences. The environment operates across time, and while some time periods may be more or less salient in terms of the influence of specific aspects of the environment, ultimately we are dealing with a long-term process when considering the role of environment in development (Pickles et al., 1989).

6

Specificity of
Environmental Action

Within the past 15 years a distinction has been made between *global* and *specific* models of environmental action. Global models of environmental action assume that "good" aspects of the environment will uniformly enhance development, whereas "bad" aspects of the environment will uniformly depress development. In contrast specificity models assume that different aspects of the environment will influence different aspects of development (Hunt, 1977; Wachs & Gruen, 1982). Within a specificity framework the goodness or badness of a specific component of the environment becomes a function of the developmental outcome that is under study. The relative recency of the global-specific distinction is, in good part, a function of methodological factors. To adequately test for the existence of environmental specificity, it is necessary to use both multidimensional environmental predictors and multidimensional outcome criteria. This design requirement was rarely met in studies of the environment done prior to 1960, which typically related differences in global social addresses (e.g., SES, institution versus home rearing) to differences in performance on global outcome measures (e.g., Bayley MDI, Binet IQ). After 1960, however, there began to be an increasing concern about the lack of "fit" between commonly used research designs and theoretical models of the environment or development. For example, Wohlwill (1973b) argued that early experience may subsume a number of distinct conditions, all of which have the potential to differentially influence development. Similarly other reviewers emphasized the problems inherent in conceptualizing intelligence as a quantitatively changing global factor (Hunt, 1961). As alternative models about the nature of environment and development became available, we began to see the emergence of research designs that could reveal the existence of environmental specificity.

Once the distinction between global and specific models of environmental action became manifest, and studies were designed that allowed an adequate test of the existence of specificity, it became possible to ask which of these two contrasting models best fit observed environment-development relations. Initial reviews on the validity of global versus specific models of environmental action came down strongly in favor of specificity (Hunt, 1979; Wachs & Gruen, 1982). Since the time of these earlier reviews additional studies have become available that can be used to determine whether initial conclusions about the existence of environmental specificity continue to hold. Two types of evidence are relevant to this question. Studies that demonstrate a relation between a given trait (t) and a specific environmental dimension (Ex), while simultaneously showing either no relation or a differential relation between another dimension of the environment (Ey) and variability on the same trait (t), can be said to offer *conditional* support for the operation of specificity. *Strong support* for the operation of specificity is seen in studies that yield a more differentiated pattern of results, namely, studies that demonstrate that environmental dimension Ex is uniquely related to variability on trait Tx but is unrelated to variability on trait Ty (or related but in a different direction), while simultaneously demonstrating that environmental dimension Ey predicts performance on trait Ty but is unrelated to performance on trait Tx (or related but in a different direction). The distinction between these two forms of support is

based on the fact that the conditional pattern may be due to
specificity; alternatively the conditional pattern also could result
from differential reliability or differential score ranges for the
different environmental predictors. Such alternative explanations
are less likely to be operating when a strong pattern of results
appears. A diagram illustrating each of the above patterns is
shown in Figure 6.1.

Current Status of the
Specificity Model

In general, evidence that has appeared over the past decade has
continued to support the validity of the environmental specificity
model. In the area of early cognitive development, *strong support*
for the operation of specificity is seen in a study by Wachs (1984),
indicating that different aspects of toddler sensorimotor development are differentially associated with distinct aspects of the
environment. For example, the development of object permanence
was shown to be inhibited by a lack of caregiver responsivity to
toddlers' vocalization, distress, or interactional behaviors. In contrast development of an understanding of causal relationships
was primarily related to caregiver support of toddlers' exploratory activities and by caregivers' teaching names and concepts to
their toddlers. In addition several studies also indicate *conditional
support* for specificity. Coates and Lewis (1984) report that early
tactile stimulation and response to distress tend to be negatively
related to early cognitive performance, whereas verbal stimulation is positively related. Bornstein (1985) notes that, while maternal teaching style in infancy predicts the child's verbal intelligence
at 4 years of age, teaching style is unrelated to general problem-
solving abilities at this age. In the most recently reported study on
this question, Bradley et al. (1990) report that the combination of
availability of play materials and parental involvement-stimulation
predicts subsequent cognitive development, whereas caregiver
acceptance of the child or use of negative control do not predict.

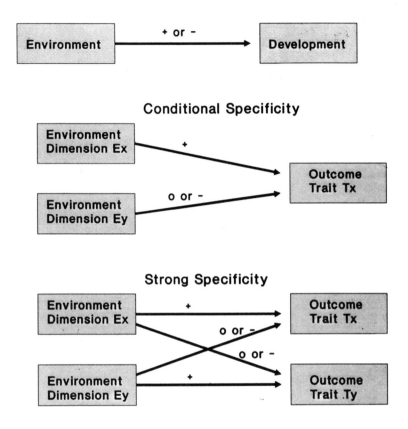

Figure 6.1. Patterns Illustrating Global and Specific Forms of Environmental Action

In the domain of language development, two studies offer *conditional support* for the operation of specificity. Golden-Meadow and Nylander (1984) have argued that the development of some aspects of language (e.g., sentence development) appears to be environmentally "resilient," in the sense that these aspects of language appear to develop almost independently of the quality of the language environment to which a child is exposed. In contrast other aspects of language (e.g., auxiliaries) appear to be environmentally "fragile," in the sense that their development depends upon the adequacy of the language environment encountered by the child. Looking at more specific environmental contributions, O'Brien (1988) has presented data indicating that, whereas the overall amount of parental speech is unrelated to children's use of object labeling, the proportion of object labels used by parents does predict children's object label fluency. *Strong support* for the operation of specificity is seen in a paper by Wachs and Chan (1986), who reported that the development of new words by the child is uniquely related to environmental indices encompassing the degree of exposure to novel stimulation (e.g., new toys in the home or changes in room decoration); in contrast the child's use of language to direct adult attention appears to be primarily related to caregiver responsivity to the child's vocalization.

Going across domains, *conditional support* for specificity is seen in two studies. Specifically Bornstein (Tamis-LeMonda & Bornstein, in press; Vibbert & Bornstein, 1989) reports that, while various indices of language development are positively related to the degree of adult teaching of the child and the degree to which the adult directs the child's attention to various aspects of the environment, these same environmental variables are unrelated to measures of the child's exploratory or play competence. Within the domain of exploratory/play competence per se, three studies offer *strong support* for the operation of specificity. Teti et al. (1988) note that mother-mediated social play at 12 months of age is associated primarily with less complex levels of infant play behavior at 18 months of age (e.g., oral or passive play), whereas mother-mediated object play predicts higher order infant play (e.g., creative play). Spangler (1989) reports that toddlers' persistence in play activities was predicted by maternal responsivity and involvement, whereas toddlers' intensity of play was positively related to the degree of symbolic or cognitive play activity the toddler was exposed to and was negatively related to the degree of exposure to social experience. Further subdividing play activities, Fiese (1990) reported that the amount of exploratory play shown by toddlers was positively related to maternal intrusive-ness and maternal use of questions, whereas symbolic play was *negatively* related to these same aspects of maternal behavior. In contrast the toddler's use of functional play was primarily related to the mother's physically directing play activities.

The potential influence of environmental specificity also has been noted in studies in the area of socioemotional development. *Strong* support for the operation of specificity is seen in studies on infant attachment. A number of studies (Belsky, Rovine, & Taylor, 1984; Isabella & Belsky, 1991; Lewis & Feiring, 1989) have pre-sented evidence indicating that avoidant attachment is a conse-quence of insensitive overstimulation by the caregiver, as opposed to resistant attachment, which is primarily related to insensitive understimulation by the caregiver.

Beyond the infancy period two studies are available, one show-ing conditional and the other strong support. Illustrating *conditional support*, Kuczynski et al. (1987) report that maternal use of direct control strategies (commands) is related to preschool children's show-ing less compliance to maternal directives, whereas maternal use of indirect control strategies (explanations) is related to greater com-pliance. For school age children and adolescents Shell, Roosa, and Eisenberg (1991) report that family stability and communication predict children's use of problem-solving coping strategies, whereas family conflict results in children's use of problem-avoiding cop-ing strategies.

In the area of child psychopathology, *conditional support* for the operation of specificity is seen in two studies. Patterson and Yoer-ger (1991) present evidence indicating greater predictability of childhood behavior disorders when specific parental factors such as monitoring or discipline are treated separately rather than being combined into a good or bad parent composite. For families at risk for depression, while family discord, divorce, and low family cohesion are all related to the development of conduct disorders, low family cohesion is also uniquely associated with the development of childhood depression (Fenderich, Warner, & Weissman, 1990).

Although the overwhelming bulk of recent evidence on envi-ronmental specificity has involved studying the child's home environment, some evidence suggests that specificity also may operate in nonhome contexts as well. One example would be the differential cognitive outcomes associated with different types of preschool programs (Miller & Bizzell, 1983). In addition the operation of specificity may not be restricted only to psychosocial rearing conditions. Several studies have provided support for the possibility that specificity also may occur when analyzing relations between nutrition, environment, and development. For example, Waber and her colleagues (Waber et al., 1981) have reported that nutritional supplementation appears primarily to influence motor development, whereas educational intervention appears primarily to influence language development. Using a sample of Egyptian toddlers Wachs et al. (in press-a) have reported that nutritional intake appears to be uniquely related to measures of toddler cognition but is unrelated to measures of toddler behavioral competence, whereas psychosocial environmental stimulation predicts behavioral competence but is unrelated to cognitive performance.

Is Specificity Real or Artifactual?

While available evidence gathered over the past 10 years has continued to support the concept of environmental specificity as a valid model of environmental action, there has been one major criticism of this model. While accepting the concept of specificity, Belsky (1990) has argued that methodological and analytic problems with most existing studies make it difficult to distinguish genuine specificity from random variability. *Methodologically* Belsky (1990) has argued that, unless specific environment-development relations are predicted *at the outset*, we run the risk of interpreting random findings as evidence for specificity of environmental action. There seems no doubt that random differences can occur as a function of methodological factors, such as inadequate measurement or restriction of range for certain environmental variables. If these types of methodological problems are operating, a pattern of results may emerge that can all too easily be incorrectly interpreted as environmental specificity. This is especially true if the results are contradictory to other research. For example, in contrast to studies noted earlier relating caregiver overstimulation to avoidant attachment, Bohlin, Hagekull, Germer, Andersson, and Lindberg (1989) report that their measure of overstimulation was associated with resistant rather than avoidant attachment. Taken in *isolation* either of these contradictory findings could be used as an example of environmental specificity. While it is possible to hypothesize, post hoc, why cross-study differences occur (e.g., Bohlin et al. suggest that intrusiveness may be highly situation specific and thus not reflect general overstimulation), such post hoc explanations should not be used to support the validity of the environmental specificity model.

In terms of *statistical criteria*, J. Belsky (personal communication) has argued that it is not sufficient to conclude that specificity exists on the basis of one correlation being significant while a second is nonsignificant (or even on the basis of one correlation being significant and positive while a second is significant and negative). Differences between correlations may reflect nothing more than random variability around a general population value. Thus the critical test for specificity is not the significance of individual correlations but whether the critical correlations are *significantly different than each other* (see Cohen & Cohen, 1983, for a discussion of the appropriate test for assessing the significance of the difference between dependent correlations; alternatively Volling & Belsky, 1991, have developed a crossover procedure that may offer an even more stringent statistical test for specificity). Unfortunately, the overwhelming majority of studies reporting evidence for specificity do not test for the significance of differences between correlations. For example, out of all the specificity studies reviewed earlier in this chapter, only one (Isabella & Belsky, 1991) makes an attempt at assessing whether different patterns of environment-development relations are really different.

Obviously the question of whether observed results reflect specificity or random relations becomes less critical if patterns of environment-development specificity can be replicated across studies. With the exception of the attachment studies, however, which were reviewed earlier (also see the cross-cultural studies described below), there have been few replication studies on the question of specificity. In good part this is because most studies in this area assess different environmental and outcome dimensions, making it difficult to compare across studies. While replication is the ultimate test of the validity of the specificity hypothesis, in the absence of replication other criteria, such as those described by Belsky, become especially important.

Fortunately some studies on specificity are available that do fulfill one or both criteria discussed above. Evidence is available from three recent studies where a priori predictions were made. While not all predictions were confirmed, those that were offer *strong support* for the operation of environmental specificity. In the first study Wachs (1987b) predicted that variability in toddler object mastery motivation would be positively related to the availability and novelty of objects in the home and negatively related to adult interference with toddler play; it was further hypothesized that toddler social mastery motivation would be negatively related to the degree of noise and crowding in the home and positively related to the amount of parental responsivity, whereas toddler social-object mastery motivation would be positively related to adult mediation of toddler's object play. While unique relations with environment appeared for social mastery motivation, these relations were not the ones hypothesized, making it difficult to claim environmental specificity for this dimension. The predicted unique relations between parental interference and toddler object mastery and between parental object mediation and toddler social-object mastery did appear, however, thus confirming the operation of environmental specificity for these outcome dimensions. While tests of the significance of differences were not made, this study was the first to actually test a priori predictions on the existence of specificity.

In the second study, based on both previous research and Piagetian theory, Jennings and Connors (1989) hypothesized that preschool children's verbal ability and play behavior should be negatively related to maternal directiveness and positively related to maternal affective tone, whereas a reverse pattern should occur for nonverbal abilities. At least for cognitive abilities the observed pattern supported the operation of specificity. Specifically, results indicated that the path between maternal affective tone and off-spring verbal ability was significant and positive, whereas maternal directiveness was unrelated to children's verbal abilities; in contrast the path between maternal directiveness and children's nonverbal abilities was negative, whereas the path between affective tone and nonverbal ability was zero order. Again, however, the significance of the differences between different environment-development patterns was not assessed.

The third study was done by Bornstein and Tamis-LeMonda (1990) using infants in the first six months of life. These authors predicted that maternal encouragement of infant social behavior should be primarily related to infant social orientation, whereas maternal encouragement of infant object play should be primarily associated with infants being oriented toward objects in the environment. In both cases the observed results fit the specificity pattern. The Bornstein and Tamis-LeMonda results are especially critical, not only because they are based on a priori prediction but also because the significance of differences in environment-development specificity patterns was *statistically tested for and confirmed*.

An even more stringent test of the validity of the specificity hypothesis is seen for those few studies that not only predict specific environment-development relations but also test for these relations *across cultures*. These studies provide critical evidence on whether specificity patterns found in one culture can be *replicated* in a different culture. Following up on their previous work on specificity, Bornstein, Azuma, Tamis-Lamonda, and Ogino (1990) predicted that maternal interactions that engage the infant in involvement with the environment should be associated with more infant exploration of the environment, whereas maternal social encouragement should be associated with more frequent infant social interactions. While tests of significance for specificity patterns were not reported, the predicted pattern was found to occur, both in Japan and in the United States.

In the second cross-cultural test for specificity, Wachs, Bishry, Sobhy, McCabe, Yunis, and Galal (1992) noted that in research studies done in North America the most salient environmental predictors of toddler vocalization involved measures of caregiver vocalization and responsivity to toddler vocalization, while the most salient predictor of amount of toddler distress was caregiver responsivity to distress. As a test of the cross-cultural generalizability of the specificity hypothesis, Wachs et al. attempted to replicate this pattern in a non-Western setting, using a sample of Egyptian toddlers who were studied from 18 to 30 months of age. Their results clearly supported the cross-cultural generalizability of the predicted specificity relation, with measures of caregiver vocalization and vocal responsivity being significantly related to toddler vocalization but unrelated to toddler distress, whereas measures of caregiver response to distress were significantly related to toddler distress but were unrelated to toddler vocalization. Even more critical, the significance of differences in these patterns was *statistically tested for and confirmed*.

CONCLUSION

Belsky has raised an important point about the necessity of distinguishing between specificity and random relations. It may well be that some of the "specificity" relations demonstrated in earlier research were simply random noise. Recent research that meets one or both of his criteria, however, continues to support the existence of specificity of environmental action. More critical, available results suggest that the specificity model of environmental action is generalizable across at least three distinct cultural groups, namely, the United States, Japan, and Egypt.

What Underlies Specificity?

One unanswered question is why environmental specificity exists. Certainly the operation of specificity would seem to violate our treasured concept of parsimony, in the sense that it would appear to be much more efficient if we had two general classes of stimulation, good or bad, which equally influence all aspects of development. There appear to be several reasons why specificity may be necessary. The first has to do with the nature of the environment and of development, while the second has to do with the organization of the central nervous system. If either development or environment were unidimensional, we should not expect specificity, because we would be dealing with global, undifferentiated variables. If both development and environment are multidimensional in nature, however, then specificity is the only logical outcome. Given that we are, in fact, dealing with predictor and outcome components, it would be rather surprising if specificity did not appear (Wachs & Gruen, 1982).

Biologically, evidence from developmental neuropsychology studies (e.g., Greenough & Juraska, 1979; Greenough et al., 1987) suggests that stimulation in specific modalities may modify specific regions of the central nervous system. Particularly for experience dependent stimulation it appears as if the impact of experience upon central nervous system development is localized in specific regions, which are involved in processing specific dimensions of information. This suggests the possibility of a neurological link and/or substrate underlying specificity of environmental action at the behavioral level.

Implications of Specificity

The concept of environmental specificity is not new. Evidence from both anthropology and human ecology has long shown parallels between specific climatological and subsistence factors and the development of specific characterological traits within a population (Moos, 1973). The extension of the specificity concept from a population level to individual development is a relatively new phenomenon, however. An increasing amount of evidence clearly supports the validity of such an extension of the specificity principle from a population to an individual level. One major implication of this extension is that it becomes increasingly inappropriate to discuss the nature of environmental influences without first specifying what aspects of the environment and what aspects of development are involved.

While there is an increasing amount of evidence supporting the validity of the environmental specificity model, it must be emphasized that the process of environmental specificity does not operate in isolation. There may well be complex relations between specificity of environmental action and other developmental processes. For example, evidence from several studies suggests the possibility that specificity of environmental action may operate across time. Yarrow et al. (1984) have presented evidence indicating that parental sensory stimulation at 6 months is primarily related to an infant's attempt to solve problems posed by objects in the environment. In contrast sensory stimulation of the infant at 12 months is primarily related to the degree to which the infant practices sensorimotor skills. Similarly Wachs (1984) has reported that the unique contributions to specific sensorimotor abilities of parental vocal stimulation and parental monitoring of infant behavior changes from the first to the second half of the second year of life. While a priori predictions about age changes were not made, in both studies the nature of the specific environment-development patterns changes across ages.

Further, while the specificity *process* does seem to operate across different cultural contexts, the operation of specificity may be useful for explaining why cross-cultural differences in behavior appear. To the extent that different cultures use different rearing patterns, from a specificity viewpoint we would expect different developmental outcomes. For example, Super (1976) has presented data indicating that the reported precocity of African infants on motor development may well be a function of highly specific training experiences endemic to specific African macro systems.

The existence of environmental specificity appears to have clear implications, both for future environmental research and for further theorizing about the nature of environmental influences. As an example of the implications of specificity for theory, let us consider the concept of within-family or nonshared environmental variance. At least on a logical level, we would expect siblings who are reared in the same family to show more similarities than differences. There is a consistent body of evidence, however, indicating that siblings reared within the same family show only low levels of resemblance in personality patterns (Plomin & Daniels, 1987; Rowe & Plomin, 1981). Based on these data, plus results from twin and adoption studies, behavior genetic theorists have argued for the existence of a *unique major aspect of environmental action,* namely, nonshared or within-family environmental variance (Plomin & Daniels, 1987; Rowe & Plomin, 1981). Nonshared environmental variance is seen as referring to those aspects of the environment that act to make individuals different (e.g., birth order, chance encounters), as opposed to those "shared" aspects of the environment that act to make individuals similar. Although the concept of nonshared environmental action has been most often applied to personality development, some researchers have pointed to the salience of this construct for understanding cognitive development as well (McCall, 1983, 1984).

The concept of nonshared environmental variance as a unique and major aspect of environmental action has been highly controversial. By emphasizing the importance of studying the environments of sibs reared within the same family, as opposed to the traditional strategy of looking at environments in different families, the concept of nonshared environment has added a new research dimension to the study of environmental influences. This is particularly true when researchers focus on sibs' perceptions of differences in their common family environment (e.g., Daniels, Dunn, Furstenberg, & Plomin, 1985). Even though this concept has been criticized on methodological grounds, as in the demonstration that nonshared environmental variance is more likely to appear when the environment is modeled but not actually measured (Rose, Kaprio, Williams, Viken, & Obrenski, 1990), what is controversial about nonshared environmental variance is not its methodological implications per se. Rather what is controversial is the assertion that the concept defines a previously undiscovered aspect of environmental influences (Hoffman, 1991; also see commentaries accompanying the 1987 Plomin & Daniels paper). I would argue that what the concept of nonshared environmental variance illustrates is not so much a new form of environmental action but the operation of environmental specificity. Within the framework of environmental specificity, if we can make the assumption that sibs residing within the same family are encountering different environments, then the predicted result would be different patterns of development for sibs. Can we make this assumption? Recent reviews (Dunn & Plomin, 1990; Hoffman, 1991) clearly illustrate that sibs *within the same family* can encounter quite distinctive environments. Given that there appear to be highly specific environment-development relations, with different environments being associated with different patterns of developmental outcomes (environmental specificity), it seems clear that the lack of resemblance in personality for sibs need not be attributed to a new aspect of environmental action, namely, non-shared environmental variance. Use of the concept of environmental specificity allows us to readily explain what at first glance seems to be a puzzling pattern of evidence in regard to sibling development, without having to add a new theoretical term to an already complex environmental system.

In regard to environmental research, as noted earlier, the existence of specificity suggests the necessity for using both multidimensional predictors and criteria when studying the nature of environmental influences. Specific environment-development relations may be lost if environmental measures are collapsed either into global factor scores or into heterogeneous composites supposed to describe the goodness or badness of an environment; similarly specific relations may be lost if performance on a number of distinct areas of development is collapsed into a single developmental score. Going beyond this general guideline, it is also essential that future studies on specificity must specify, a priori, the nature of environment-development relations that are expected. Predicted or replicated relations can be seen as evidence for specificity and can be used as a building block for future theory. Specific environment-development relations that are neither predicted nor replicable would suggest the operation of random variability rather than specificity.

Specific specificity predictions can be based either on previous research or on theoretical statements defining which aspects of the environment should be relevant to a specific developmental outcome. In the absence of prior research or specific theories, one approach to making relatively specific predictions about environment-development patterns could involve the use of a task analysis procedure. Task analysis procedures have been used by cognitive psychologists (Kail & Bisanz, 1982) to break down complex cognitive tasks into a series of specific stages. A similar procedure has been employed by social development researchers (Dodge, Pettit, McCluskey, & Brown, 1980) to decompose children's social competence into a series of specific steps, including coding and interpretation of social stimuli and generation of responses based on the child's interpretation of social stimuli. Rather than attempting to relate specific aspects of the environment to complex cognitive processes such as object permanence or to complex social processes such as compliance, a task analysis approach would necessitate developing a relatively specific set of steps that the child would need to go through to show object permanence or to display compliance. Breaking complex processes down into a relatively discrete set of steps may make it easier for researchers to hypothesize logical environmental precursors of these specific steps and hence make relatively specific environment-development predictions. Being able to make relatively precise predictions about specific environment-development relations will not only tell us more about the nature of specificity per se but will also enable us to develop more precise theories about the nature of environmental influences in general.

Organism-Environment Covariance

Although we have been relatively successful in identifying specific determinants that relate to variability in development (e.g., genes, environment, nutrition), for the most part these factors have been studied in isolation. While it may make good scientific sense to study the influence of a single determinant in isolation, does this strategy reflect the nature of reality? Is variability in development nothing more than the sum of individual determinants considered in isolation? Elsewhere (Wachs, in press) I have argued that there is evidence available that suggests at least two ways in which different determinants interface with each other to produce developmental variability. One, *covariation* (correlation) among different determinants, will be considered in this chapter. The second, *interaction* among different determinants, will be considered in the next chapter.

The idea that variability in behavior and development is governed by a system of covarying environmental and nonenvironmental influences has been repeatedly suggested (e.g., Horowitz, 1987; Plomin, Loehlin, & DeFries, 1985; Wachs, 1983). Simply stating that environment covaries with other developmental determinants is

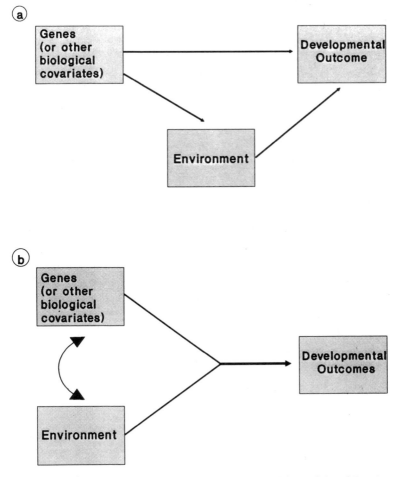

Figure 7.1. Deterministic (a) Versus Probabilistic (b) Models of Passive
Organism-Environment Covariance

various stressful life events (Garmezy, 1987) and to have parents
who are more likely to be authoritarian and stress conformity
(Schaefer & Edgerton, 1985).

While there appear to be no clear examples of reactive or active
environment-environment covariance, there have been several

hypotheses about how these might occur. Liben's (1981) suggestion that environments with little differentiation or organization may force individuals to impose their own structure on the environment may be the environmental analogue to reactive covariance. The analogue for active covariance may be seen in the suggestion by Bateson (1983) that imprinting may act to restrict social preferences to individuals who are familiar, thus preventing further social experience from changing social preferences. The potential operation of environment-environment covariance in any form reinforces the conclusion drawn at the end of Chapter 4; namely, that, to understand the nature of environmental influences, it is necessary to focus on more than just one level of the environment.

BIOLOGICAL-ENVIRONMENTAL COVARIATION

Gene-environment covariance. Adoption study research nicely illustrates the operation of passive gene-environment covariance. The logic underlying adoption studies is that in adopted families caregivers only pass along environments to their children, whereas for nonadopted children their parents pass along genes, environment, and the covariance between genes and environment. To the extent that passive gene-environment covariance is operating, we would expect to find higher correlations between environment and developmental outcomes for biological children (Plomin et al., 1985). Available evidence generally conforms to this predicted pattern of stronger environment-development correlations in nonadoptive than in adoptive families (Plomin & Bergeman, 1991; Plomin et al., 1985).

Another example of passive gene-environment covariance comes from high risk studies, involving children at developmental risk due to a history of parental psychopathology. If passive gene-environment covariance does exist, we would expect to see evidence for disturbed parent-child relations in these types of families. In general this is the pattern that emerges. Specifically parental mental illness (either schizophrenia or depression) is found to covary with more *marital discord* and *family stress* (Billings & Moos,

1983; Emery, Weintraub, & Neale, 1982), *poorer social support* networks (Walker & Emory, 1983), a greater likelihood that children in these families will be *maltreated* (Walker, Downey, & Bergman, 1989), poorer organization of the home environment, and less emphasis on developing offspring independence (Billings & Moos, 1983). While high risk studies do suggest the operation of gene-environment covariance, one problem is that risk populations are not genetically homogeneous. Thus not all children in a risk population will have a genetic liability for the disorder under study (Erlenmeyer-Kimling & Cornblatt, 1987). Under these conditions it becomes unclear whether more disturbed parent-child relations in high risk families reflect covariance or a cross-generational environmental transmission effect. The use of biological or behavioral markers of genetic risk may be one way of identifying children with a real genetic liability for a disorder (Erlenmeyer-Kimling & Cornblatt, 1987).

Environment-biomedical covariance. As a possible example of active covariance, Breitmayer and Ramey (1986) have noted that children with a history of biomedical risk may be less motivated to seek out cognitively stimulating activities than children without such a history. While this example is intriguing, currently most examples of biomedical-environment covariance involve passive or reactive covariance. At the level of the family Horowitz (1987) has hypothesized that a handicapped child may be more likely to influence family functioning than a nonhandicapped child. Confirming this hypothesis, Nihira et al. (1987) report greater congruence of parental beliefs and attitudes between Japanese and American parents of retarded children than between these parents and parents of normal children within their own culture. This congruence suggests that the impact of a retarded child upon the family may transcend even culturally mediated rearing patterns and beliefs.

In terms of specific covariates there are a number of studies indicating that children having more biomedical problems are more likely to encounter supportive environments, defined in terms of either greater caregiver responsivity (Cohen, Parmelee, Sigman, & Beckwith, 1982) or more positive attitudes by caregivers (Greenberg & Crnic, 1988). Cross-culturally, in both Kenya and Egypt, toddlers having higher rates of illness tended to receive more physical contact stimulation from their caregivers (Wachs et al., in press-b).

In contrast, evidence also is available suggesting that more biomedical problems covary with less adequate environments. Specifically, available evidence indicates that the more economically disadvantaged the family, the more likely children in these families are to be at risk for biomedical problems (Joffe, 1982). These relations may hold even within relatively restricted demographic groups. For example, Douglas (1975) has reported that, in working-class families in Britain, children with repeated hospital admissions were more likely to come from large families and to have parents unconcerned with their schoolwork.

The above findings may seem paradoxical, in the sense that biomedical risk factors seem to be associated simultaneously with both negative and positive aspects of the environment. The paradox is easily resolved when we remember that the structure of the environment operates at different levels. It seems clear that at the macrosystem level there is a covariance between biomedical risk and environmental disadvantage, with children living in disadvantaged circumstances being more likely to be at risk for biomedical problems. At the microsystem level, however, caregivers do attempt to provide more supportive stimulation to the ill child. Lending further support to the concept of the environment as a multidimensional system, whether the caregivers of physically ill children actually are able to function in this way may depend upon their environmental context. Specifically, when there is a high level of macrosystem disadvantage, even the most well-motivated caregiver may be unable to provide for the special needs of a child with biomedical problems (Joffe, 1982). For example, evidence indicates that, in extremely impoverished or stressful environments, parents often are not able to adapt their activities sufficiently to take care of the emotional well-being or needs of the physically ill child (Pollitt, 1983).

Covariance between nutrition and environment. In general, available evidence indicates that less adequate nutritional status passively covaries with less adequate environments, such as higher levels of crowding (Powell & Grantham-McGregor, 1985) or lower quality caregiver-child relations (Cravioto & DeLicarde, 1972; Engle, 1990). This seemingly straightforward relation is moderated, however, by a variety of other factors. For example, evidence is available suggesting that the nature of the covariance between nutritional intake and the quality of the psychosocial rearing environment may be mediated by higher order macroenvironmental factors. Specifically Wachs et al. (in press-b) report that poorly fed Kenyan toddlers were more likely to be responded to by their caregivers; in contrast poorly fed Egyptian toddlers were less likely to be responded to. Given that the availability of food was greater in Egypt, it may be that Kenyan caregivers were attempting to compensate for the poor nutritional intake of their children by being more responsive; poorly fed Egyptian toddlers may come from families that are deficient in providing both nutrition and psychosocial rearing. These data suggest that the covariance between nutrition and environment also must be viewed as part of a system and not in isolation.

In addition to passive covariance, we also have the possibility of reactive nutrition-environment covariance, wherein more adequately nourished children, being more active (Torun, 1991), may be better able to elicit developmentally facilitative behaviors from their caregivers (Chavez & Martinez, 1984). Again, however, reactive nutrition-environment covariance must be considered part of a system, given that this pattern appears to be moderated by a variety of other factors such as sex and age of the child (Torun, 1991) and the nutritional status of the caregiver (Cravito & DeLicarde, 1972; Wachs, 1991c).

CHILD-ENVIRONMENT COVARIANCE

Temperament-environment covariance. One possible example of an individual child characteristic that may serve to elicit specific types of caregiver reactions (reactive covariance) is temperament: the biologically rooted behavioral style of the child (Bates, 1989). Supporting this idea, a number of studies have reported that young children having difficult temperaments (highly intense, easily excited, negative mood states) are more likely to elicit negative responses (Gordon, 1981) or reduced teaching efforts from caregivers (Maccoby, Snow, & Jacklin, 1984). Indeed some evidence even suggests a possible link between an infant's difficult temperaments and maternal depression (Cutrona & Troutman, 1986). Unfortunately, available reviews of the temperament-caregiver literature report that it is just as easy to find other methodologically sound studies that do not demonstrate associations between child temperament and caregiver behavior patterns (Bates, 1989; Crockenberg, 1986; Slabach, Morrow, & Wachs, 1991).

There are a variety of potential reasons why more consistent results are not found, all of which again emphasize the need to consider organism-environment covariance within a systems framework. Potential reasons include the following:

(1) Individual caregiver characteristics may moderate relations between child temperament and parent behavior. For example, Lounsbury and Bates (1982) have presented evidence indicating that experienced mothers are less upset by cries of infants with difficult temperament than are less experienced mothers. Hubert and Wachs (1985) report that child characteristics that cause parents to label their infant as difficult may vary from parent to parent, depending upon the parent's preference for individual infant behaviors. Parental beliefs about their ability to cope with their infant's behavior also have been reported as moderating parental reactivity to infant difficultness (Cutrona & Troutman, 1986). Similarly, the teacher's temperamental style may influence his or her tolerance for specific child temperament characteristics (Keogh, 1986b).

(2) Nontemperamental characteristics of the child may moderate relations between child temperament and caregiver behavior. Gordon (1983) has presented evidence suggesting that mothers may be less tolerant of difficult behavior from daughters than from sons. Keogh

(1986b) has reported that nonhandicapped children with easy temperament patterns and handicapped children with difficult temperament patterns get more teacher interaction than normal difficult or handicapped easy children.

(3) Another possibility is the mediation of temperament caregiver behavior relations by environmental context. Certainly cultures differ in terms of what child behavioral characteristics are labeled as difficult (Super & Harkness, 1986). Other evidence suggests that caregivers in high stress environments may be particularly intolerant of difficult temperament children (Hannan & Luster, 1990; Rutter, 1979). As might be expected parents with adequate levels of social support who have irritable, difficult infants are more likely to be responsive (Crockenberg, 1981) and sensitive (Crockenberg & McClusky, 1986) to these infants than are parents with low levels of support.

(4) The degree of covariance between child temperament and caregiver behaviors may depend on the stability of child temperament over time. In this regard Lee and Bates (1985) reported that children rated as temperamentally difficult at 6 and 13 months were more likely to have mothers who used power assertion discipline techniques at 24 months of age. When Lee and Bates partialed out 24-month temperamental difficultness, however, the significant relations between early temperament and later maternal use of power assertion techniques dropped to nonsignificance, suggesting that it is the continuity of difficult temperament between 6 and 24 months of age that relates to caregiver behavior patterns.

Overall the above pattern of results suggests that, while there may well be a covariance between individual child temperament and the child's rearing environment, the relation is not a simple one. Temperament may be a necessary but not a sufficient predictor of variability in caregiver behavioral patterns. Rather than continuing to assume a direct relation between temperament and caregiver behavior patterns, what may be necessary is the development of systems models integrating both temperamental and nontemperamental influences, as these jointly predict caregiver behaviors. Potential system parameters that could moderate the covariance between child temperament and environment include nontemperamental individual child characteristics (e.g., age, sex), caregiver characteristics, and environmental context.

Attachment history. Biologically based variations in child characteristics are not the only ones that may covary with environment. One environmentally based child characteristic that also may reactively covary with rearing environment is the child's attachment history. The strongest evidence for environment-attachment covariance are studies that follow up children with different attachment histories. In a recent summary of this research Sroufe and Egeland (1991) report that nursery school teachers generally were warm and age appropriate in their relations with securely attached children. In contrast teachers were warm and controlling toward children with anxious resistant attachments, treating these children as if they were younger than their chronological age. For children with avoidant attachment histories, teachers not only were not nurturant but at times even displayed anger toward these children. What is particularly interesting is that the patterns of behavior shown by teachers toward children with different attachment histories are strikingly reminiscent of the types of mother-infant relationship patterns thought to produce differences in attachment in the first place. In addition to attachment history reactively covarying with adult-child relations, Sroufe and Egeland (1991) also report that children with histories of insecure attachment in infancy were more likely to be socially isolated or rejected by school age peers than children whose earlier attachment had been rated as secure.

There are a variety of potential reasons why children with different attachment histories are more likely to experience different environments. Based on an analysis of differences in play behavior among children with different attachment histories, Slade (1987) suggests that securely attached children are more able to tolerate frustration and delay gratification; as a result these children may be more adept at soliciting support from adults in their environment than are less securely attached children. Sroufe (1983) has suggested that early attachment relations define expectancies the child has about other people; these expectancies in turn relate to how the child approaches and copes with the environment, which in turn influences how others treat the child later in life. Sroufe and Fleeson (1986) also have suggested that children attempt to reestablish relationships that are congruent with past relationships. Thus a rejected child may misbehave until a familiar relationship pattern reappears, involving rejection from others in the environment. This latter possibility suggests the operation of active covariance, in that children with different attachment histories may seek out specific environmental "niches."

Cognitive status. Kindermann and Skinner (1988) hypothesize that, in a given culture, there are certain developmental achievements that are seen as central to a child's competence in that culture. While the nature of the central tasks varies across cultures, the degree of competence the child displays in these tasks is viewed as structuring the nature of caregiver behavior patterns toward the child (reactive covariance). Supporting this hypothesis, several studies have reported covariance between child intelligence and caregiver behavior. Bornstein (1985) has reported evidence indicating that the level of infant attentional capacities at 4 months of age relates to the degree to which mothers encourage their children's attention at 12 months of age. Heckhausen (1987a, 1987b) has presented evidence suggesting that mothers adapt their behaviors to their infants, based on their perception of the infant's level of development. Smith and Hagen (1984) postulate a similar process, noting that, while there were no differences in the environments of 7-month-old normal and mentally retarded Down syndrome infants, a number of differences had appeared by 17 months of age. As a possible explanation Smith and Hagen report that Down syndrome mothers typically felt that their infant's behavior was age appropriate at 7 months of age; however, mothers clearly were concerned about developmental delays that had occurred by 17 months of age. It is also worth noting that processes underlying cognitive reactive covariance may not be restricted to mothers. As reported by Teti et al. (1988), the object skills displayed by younger sibs at 12 months of age relate to the degree their older sibs vocally interact with them at 18 months of age.

Child behavior patterns. Control systems theory is based on the assumption that caregivers have upper and lower levels of tolerance for the intensity and appropriateness of children's behavior (Bell & Chapman, 1986). When children's behavior becomes overly intense or inappropriate, the caregiver is more likely to react by attempting to bring the child's behavior under control, using such techniques as redirection or punishment. When the child's behavior is overly passive or inactive, the caregiver is more likely to react by trying to raise the level of the child's behavior, through rewarding or prompting the child. Thus control systems theory predicts a specific pattern of reactive child behavior—caregiver behavior covariance. In a review of both experimental and longitudinal studies, Bell and Chapman (1986) report that children's dependence is more likely to result in parental directiveness than is child independence. Similarly overaggressive or overactive children appear more likely to elicit parental directiveness or power assertion than less active or less aggressive children.

Evidence from other domains also tends to show caregiver-child relational patterns that are consistent with predictions derived from control systems theory. For example, Martin (1981) reports a negative relation between the degree of child compliance at 2 years of age and the degree of maternal manipulativeness when the child is 3½ years. Similarly, the greater directiveness and negativeness shown by parents of hyperactive children tends to be reduced after the children become more compliant, following treatment with stimulant medication (Campbell, in press).

While there are a number of studies indicating that the intensity and appropriateness of children's behavior can relate to subsequent caregiver behavior, not all studies show the pattern predicted by control systems theory. Bell and Chapman (1986) themselves note that available results are not entirely consistent with control systems theory when the behavior under study involves measures of children's sociability. In more recent research Pianta, Sroufe, and Egeland (1989) report that boys who were less social received more sensitive care than expected from their mothers, perhaps as a means of compensating for the child's lack of sociability. In contrast to what would be predicted from control systems theory, however, a similar result does not occur for girls; rather, girls who were characterized as more controlling and persistent were more likely to experience sensitive care from their caregivers.

In analyzing the reasons for inconsistencies between predicted and observed relations, Bell and Chapman (1986) suggest that both ecological contextual factors (e.g., subcultural beliefs or expectations) and individual characteristics may moderate parents' reactions to children's behavior patterns. For example, in a lower-class family with a large number of children, the mother may be overwhelmed and simply give up attempting to elicit more intense levels of behavior from a very passive child. Alternatively, mothers who do not believe that they have control of their children's behavior may give up trying to use either upper- or lower-limit control behaviors.

One potential individual characteristic that could act as a moderator is caregiver attachment history. Crowell and Feldman (1988) have noted that the mother's own attachment history is related to her reaction to her child's behavior, with more optimal reactivity by the mother being associated with the mother's having a history of secure attachment.

In terms of contextual factors, Bradley et al. (1990) report that, while lower-class parents tend to tailor their reactions to their toddlers' degree of reactivity (as would be predicted by control systems theory), middle-class parents appear to be more proactive, attempting to influence their children without necessarily waiting for a specific reaction pattern to appear. At the microsystem level the family's organization or goals also may serve to influence the nature and degree of parent-child covariance. In a study of parents' relations with their retarded children, Mink and Nihira (1986) have noted that a high degree of covariance is found for families that stress learning, with parents characteristically attempting to alter the family situation to fit specific needs of their child. In contrast, in families that are characterized by a hierarchical family structure and an emphasis on enforcement of family rules, there is a lower probability that child characteristics will influence patterns of family relations. In families where there is an emphasis on nonfamily activities, a complex bidirectional pattern is shown, with a covariance between child characteristics and family behavior being shown for some child characteristics (child's level of basic maintenance skills) but not for others (child's level of social adjustment).

Moderators of individual parental reactivity to child behavior can be biological as well as psychological. Thus recent evidence has indicated that mothers who have low levels of vitamin B6 intake are less responsive to their infants' vocalization and distress signals than are mothers with adequate B6 intake (McCullough et al., 1990). Moderators may also involve the chronosystem. For example, Wahler and Dumas (1989) conclude that caregivers tend to classify current child behavior patterns along the lines of their past interchanges with their child, whether good or bad.

Conclusions

From the information reviewed in this chapter, it seems clear that the psychosocial environment of the child covaries with a variety of biobehavioral, environmental, and individual child characteristics, many of which also have the potential to influence the development of the child. What we also see is that this covariance process *cannot be considered in isolation.* Rather covariance must be viewed as part of a more complex general system, encompassing not only child characteristics or child biobehavioral status but also caregiver beliefs, caregiver characteristics, and contextual influences. What are the implications of organism-environment covariance for understanding the nature of environmental influences upon development? There appear to be three major implications that need to be considered.

COVARIANCE AND THE CONCEPT OF RISK

Disadvantaged psychosocial rearing environments may covary with a host of other factors, such as genetic risk, inadequate nutrition, problematic child characteristics, and greater environmental stress. Available evidence clearly indicates that exposure to multiple risks results in more adverse developmental consequences than exposure to single risk factors. For example, Sameroff, Seifer, Barocas, Zax, and Greenspan (1987) report a linear relation between the number of genetic and environmental risk factors children encounter and children's IQ, with IQ scores declining as much as 30 points as the number of risk factors increases. Similar patterns also have been seen in studies by Bradley et al. (1989), Evans et al. (in press), and Meyer-Probst, Rosler, and Teichmann (1983). Further, this pattern is not restricted to the area of cognitive development; similar findings also are seen for social-emotional development (Belsky & Rovine, 1988; Billings & Moos, 1983). In interpreting these data what is essential to keep in mind is that simultaneous exposure to multiple risks is not likely to occur randomly. Rather, as shown throughout this chapter, risk factors covary, so that exposure to a specific risk factor increases the probability of exposure to other risk factors. Exposure to more risk factors in turn increases the probability of less adequate developmental outcomes.

The degree of developmental deficit associated with multiple risk factors may increase even more as children continue to be exposed to covarying multiple risks. An excellent example of this "cumulative deficit" process is seen in a report by Saco-Pollitt, Pollitt, and Greenfield (1985). Comparing changes in cognitive performance of advantaged children in the United States versus disadvantaged children in Central America, these authors report that, as U.S. children got older, their cognitive performance increased; in contrast, in Central America there was little evidence for increases in cognitive performance across time, indicating that children exposed to covarying multiple risks were less likely to show the benefits of previous learning as they grew older. As E. Werner (personal communication) has pointed out, the nature of the covariance process underlying these types of cumulative deficits remains unclear. Specifically Werner emphasizes the importance of distinguishing between *simultaneous covariance* (the impact of simultaneously occurring risk factors accelerating across time) versus *sequential covariance,* as when the occurrence of one risk factor leads to the occurrence of a second risk factor (e.g., undernourished children become more susceptible to illness).

Children's development is not the only thing likely to be compromised when different risk factors covary. Also compromised is the clinician's ability to provide effective intervention for these children. Comparing situations where treatment gains for children were maintained across time versus situations where treatment gains were lost, Wahler and Dumas (1989) conclude that treatment gains are less likely to be maintained in situations where families are exposed to multiple covarying stressors. These treatment losses may occur because covarying multiple stressors impede the caregiver's ability to adequately monitor either the child's behavior or the impact of his or her own behavior upon the child (Wahler & Dumas, 1989).

While the overall pattern of evidence strongly suggests that covariance among multiple risk factors is more likely to lead to developmental deficits, one important qualification must be considered. Given the evidence cited in Chapter 4, it is not surprising to find that the "riskiness" of a specific factor may be moderated by the cultural context within which this factor operates (Super & Harkness, 1986). Factors that are risks in one context may be nonrisks or even beneficial in a second context. For example, the temperamental dimension of low rhythmicity, which is a child characteristic that helps to define difficult temperament in Western societies, is found to be relatively unimportant for children's behavioral development (i.e., not a risk factor) in both Africa (Super & Harkness, 1986) and India (Malhotra, 1989). The difference may be due to a lessened emphasis on rigid time schedules in these cultures. Similarly, while difficult temperament is a primary risk factor (predicts behavior problems) in a middle-class sample, the primary temperamental risk factor for lower-class Puerto Rican children is activity level (Korn & Gannon, 1983).

Differential riskiness goes beyond just temperament. The use of multiple caregivers is seen as a risk factor in the United States but appears to be more of a protective factor in Colombia (Super & Harkness, 1986). Similarly illness may be a much greater risk factor in countries where there is chronic malnutrition than in countries where nutritional intake is adequate (Pollitt, 1983). Indeed the argument has even been made that, under some circumstances, reduced dietary intake may not necessarily be considered a risk (for detailed discussion of this point, see Scrimshaw & Young, 1989). Thus, while it seems clear that, the greater the number of covarying risks, the more likely behavioral deficits are to occur, we must not assume that there is a standard set of risks. Rather the definition of a major risk factor may vary from context to context. Put in terms of our discussion in Chapter 4, macrosystems may mediate not only microenvironments but also what constitutes a risk factor at the microenvironment level.

THE FIT OF ENVIRONMENT TO CHILD

Clearly there are many situations where "undesirable" child characteristics covary with "less adequate" rearing environments. There is also the possibility, however, that, by the luck of the draw, existing child characteristics may match parent expectations or needs (see McCall, 1983, for a discussion of the role of chance in the process of environmental influences). In this type of situation the child may be more likely to benefit developmentally (Goodnow, 1988). Situations where child characteristics are congruent with environmental demands or expectations are examples of the concept of "goodness of fit" (Chess & Thomas, 1989). There are a number of case study or small sample research examples of the operation of goodness of fit (Thomas & Chess, 1991). In the most dramatic of these DeVries (1984) reports that, among the Masai in east Africa, infants with difficult temperament were more likely to survive a serious drought than were temperamentally easy infants.

In part this is because, within this culture, infants who were fussy and irritable were more likely to be preferred and fed by caregivers, based on the fact that these infants were viewed as fitting the cultural preference for a "warrior." While the concept of fit is intuitively appealing, and while there are a number of examples describing the operation of goodness of fit, for the most part research studies investigating the impact of fit have yielded inconsistent results. For example, Martin (1981) has reported that the occurrence of child sociability is greater when mothers who emphasize physical contact stimulation have children who enjoy physical contact. This relation appears to hold only for boys, however. Similarly, studies of the individual's perception of the degree of fit between what he or she prefers and what the environment offers either show little (Fraser & Fisher, 1983) or no influence of fit (Greenberger, Steinberger, & Vaux, 1982; for reviews on fit, see Lerner et al., 1986; Slabach et al., 1991).

The strongest positive evidence for the operation of goodness of fit phenomena comes from studies of children growing up under multiple risk conditions who managed to successfully cope with covarying multiple stresses. Across four major studies (Hetherington, 1989; Masten, 1989; Radke-Yarrow & Sherman, in press; Werner & Smith, 1982), those children who successfully coped with a variety of extreme life stressors were consistently found to have certain individual personality characteristics. These included sociability or social interpersonal skills, cognitive competence, and a sense of self-esteem. These individual child qualities appear to be ones that were more likely to be appreciated by individual caregivers (i.e., reactive covariance). The fact that these individual child characteristics were more appreciated by caregivers had two possible consequences. First, the child who matched the caregiver's need (fit) was more likely to receive whatever psychological assets (e.g., warmth, support) were available in the family, even if these assets were minimal (Radke-Yarrow & Sherman, in press). Second, children with these characteristics were less likely to receive parental criticism, anger, or abuse (Hetherington, 1989).

This latter group of studies, while supporting the relevance of development of a positive covariance or fit between parent needs and child characteristics, also suggests a limitation on the operation of positive fit. Specifically, the consistency of evidence in stress studies, as opposed to the inconsistency of evidence in other studies, may mean that positive fit operates as an example of what Rutter (1983b) has called an "environmental buffer." That is, the impact of positive fit may become obvious only under situations where the child or family is under stress. Under normal nonstress circumstances the positive covariance between child characteristics and caregiver needs, while operative, may not exert a major influence upon observable variability in children's development. In contrast, under stress, this positive covariance could function as a major protective influence for the child.

COVARIANCE AND THE ATTRIBUTION OF CAUSALITY

To the extent that environment covaries with biological, individual, and ecological characteristics, it is important to avoid the temptation of attributing variability in development just to environmental influences. For example, when a child is at risk for developmental deficits due to growing up in a poverty environment, it is probably impossible to separate out whether observed developmental deficits are a function of the environment or the fact that characteristics of the child's environment covary with nutritional, biomedical, or genetic risks, each of which can also influence development. In all fairness, however, the same statement could also be made in regard to attributing causality to genetic, nutritional, temperamental, or biomedical factors, given that these covary not only with each other but also with the quality of the child's psychosocial rearing environment. The fact that we have an interrelated system of causal influences means that we must avoid attributing developmental outcomes to a specific level or cause. Rather, because we have a covarying system of influences, variation in development is more properly attributed to the interrelation of variables in the system. Indeed the evidence presented in this chapter could be used to argue for the possibility that a single risk factor, unless of undue severity (e.g., severe central nervous system injury, severe prolonged malnutrition), may have few long-term developmental consequences (Sameroff et al., 1987).

Clearly it is very tempting for researchers to investigate a single, preferred determinant and assign variability in behavioral development primarily to this determinant. What I am suggesting is that we may be going down the wrong path by assuming that development is due solely to the rearing environment, or to genetics, or to nutrition, or to individual characteristics, or to the broader ecological context. Rather the major influence upon development may lie in the degree of directional consistency and in the extent of covariance among these individual determinants, both at a given point in time and across time as well (Gottlieb, 1991; Wachs, in press).

For a long time behavioral researchers have been warned against attributing causality to correlational data. The converse, however, as expressed here, suggests that the correlation (among determinants) is the causation (see Figure 7.1, part b). The impact of this message both for future developmental research and for intervention strategies will be considered in Chapter 9.

Organism-Environment
Interaction

As noted at the beginning of Chapter 7, there appear to be two ways in which different determinants may interrelate to produce developmental variability. The first, covariation among different determinants, was considered in the previous chapter. In this chapter we will look at the *interaction* among determinants. In one sense organism-environment interaction may be seen as the mirror image of organism-environment covariance. Covariance occurs when children with different characteristics elicit different types of reactions from their environment. In contrast interaction occurs when there is differential reactivity by different individuals to *objectively similar environmental stimulation* (Wachs & Plomin, 1991).

The concept of organism-environment interaction is an intuitively appealing one that, in one form or another, has been with us for many centuries. The phrase in Isaiah (6:9) "hear ye indeed but understand not; and see ye indeed but perceive not" can be viewed as the biblical precursor for the modern-day distinction between objective environment (stimuli that impinge upon the individual) versus experience (stimuli that actually influence behavior).

The Greek philosopher Aristotle was one of the first to promulgate the importance of tailoring education to the child's individual characteristics (*Nicomachean Ethics,* Book 10, chap. 9: McKeon, 1941).

In studies involving lower organisms there is ample evidence for the operation of organism-environment interaction (for reviews, see Freedman, 1974; Fuller, 1967; Henderson, 1980; Hinde & Stevenson-Hinde, 1973). For example, in a very well-controlled study, monkeys from three different macaque strains were reared in total social isolation for the first six months of life. When tested on various measures of social and behavioral competence after the isolation experience, three different patterns of outcomes were observed: Rhesus macaques showed clear deficiencies in individual behavior, social interaction, and object exploration; crab-eating macaques showed essentially normal social behavior and object exploration but were deviant in patterns of individual behavior; pigtailed macaques showed normal individual behavior and object exploration but were clearly deficient socially (Sackett, Ruppenthal, Fahrenbruch, Holm, & Greenough, 1981).

Ample evidence for the existence of organism-environment interaction is also found in human biobehavioral studies. Differential reactivity in response to specific drugs has been consistently reported and related to a variety of individual characteristics, such as age, sex, and emotional state (for a brief review, see Neims, 1986). For example, hyperactive children who were chronically high in anxiety showed significantly poorer responsivity to stimulant medication than did low-anxiety hyperactive children (Pliszka, 1989). In terms of diet, individual sensitivities in reaction to specific nutrients will determine whether intake of a specific nutrient is helpful, harmful, or irrelevant to individual functioning (for brief reviews on this topic, see Beaton, 1986; Rutter & Pickles, 1991). For example, the relation of salt intake to blood pressure is not a constant but depends on the individual's degree of sensitivity to diets that are high in salt (Kawasaki, Delea, Barter, & Smith, 1978).

Supporting the results of infrahuman studies and human biobehavioral studies, there is ample *clinical* evidence indicating generalized differences in reactivity by individual children. In one of the earliest clinical demonstrations of organism-environment interaction, Bergman and Escalona (1949) identified five children who seemed to have an unusually high sensitivity to sensory stimulation. That is, variation in stimulation that would go unnoticed by ordinary children was reported as having an unusually strong impact upon these highly reactive children. More recently researchers have concentrated on the development of "resilient" children, namely, children who appear to suffer relatively few long-term effects of exposure to continued stress during childhood (Langmeier & Matejcek, 1975; Radke-Yarrow & Sherman, in press; Rutter, 1985c; Werner & Smith, 1982). Even for children being reared in totally inhuman environments, such as concentration camps, there appear to be individual differences in the degree to which surviving children are adversely affected (Moskovitz, 1985).

Methodologically the results from clinical research studies are not easy to integrate with the findings cited earlier involving infrahuman populations or biobehavioral parameters. Infrahuman or biobehavioral studies typically involve the study of interactions between well-defined individual characteristics and experimental (environmental) manipulations. Clinical studies, while consistent with the overall hypothesis that there are individual differences in reactivity, typically involve global environmental and individual characteristics. While there are a number of studies at the human level that have looked for interactions between relatively specific individual and environmental characteristics, conceptual, statistical, and methodological problems all have hampered our ability to draw firm conclusions from this body of data (see below).

Given these problems it is not surprising that we often have difficulty demonstrating reliable differential reactivity between *specific individual characteristics* and *specific types of environmental stimulation* in human behavioral studies (Cronbach & Snow, 1977; Plomin & Daniels, 1984). The focus in this chapter will be on areas where available evidence suggests some degree of consistency of findings, in terms of identifying *specific organism-environment interactions*. Examples of areas where inconsistent findings are the rule rather than the exception are shown in Table 8.1.

TABLE 8.1 Inconsistent Findings on Organism-Environment
Interaction

	Proposed Interactions
Areas	*Findings*
1. Are children with neonatal problems more sensitive to environmental influence than children without neonatal problems?	*Yes:* Greenberg & Crnic (1988); Siegel (1982).
	No: Breitmayer & Ramey (1986). May be more sensitive to negative stimulation and less sensitive to positive stimulation: Field (1981).
	May Depend on sample characteristics: Oehler, Eckerman, & Wilson (1988).
2. Do risk children require or process different types of stimulation than nonrisk children?	*Yes:* Frankel, Simmons, Fichtner, Freeman (1984); Gersten (1983); Magnusson (1988).
	Complex group by procedure or age interactions: Brimer & Levine (1983); Schmidt, Solant, & Bridger (1985); Smith & Hagen (1984); Zentall & Shaw (1980).
3. Does stress have a greater negative impact on males than on females?	*Yes:* Compas, Slaven, Wagner, & VanNatta (1986).
	No: Simmons, Burgeson, Ford, & Blyth (1987); Vaux & Ruggerio (1983).
	Complex sex by age or outcome interactions: Burke, Moccia, Borus, & Burns (1986); Sroufe & Egeland (1991); Werner & Smith (1982).
4. Are boys more adversely affected by day care than girls?	*Yes:* Belsky & Rovine (1988); Chase-Lansdale & Owen (1987); Egeland & Farber (1984); Howes & Olenick (1986).
	No: Egeland & Farber (1984); Howes & Stewart (1987).
	Complex sex by outcome measure interaction: Hock & Clinger (1980).

5. Are boys more influenced by classroom characteristics than girls?

Yes: Trickett (1983); Wade (1981).

No: Thomas & Berk (1981).

Complex sex by classroom characteristics interaction: Miller & Bizzell (1983).

6. Are males more adversely influenced by environmental noise than females?

Yes: Wachs (1979, 1987b).

No: Hambrick-Dixon (1986, 1988).

Males less reactive: Christie & Glickman (1980).

7. Are female infants more reactive to caregiver stimulation?

Yes: Gunnar & Donahue (1980).

No: Bradley et al. (1989, 1990); Siegel (1982); Wachs (1987b).

Complex sex by outcome measure or sex by stimulation interaction: Frankel & Bates (1990); Wachs, Bishry, Sobhy, & McCabe (1991).

8. Are there sex differences in the reactivity of older children to caregiver stimulation?

Boys are more sensitive: Crouter, MacDermid, McHale, & Perry-Jenkins (1990); Walker, Downey, & Bergman (1989).

Girls are more sensitive: Cummings, Pellegrini, Notarius, & Cummings (1989); Garmezy (1987); Hess & McDevitt (1984); Kogan (1983).

No sex differences: Cummings (1987); Cummings, Zahn-Waxler, & Radke-Yarrow (1981); Dornbusch, Ritter, Leiderman, Roberts, & Fraleigh (1987); Parpal & Maccoby (1985); Pettit, Dodge, & Brown (1988).

Complex higher order interactions between child sex and age: Vaughn, Block & Block (1988); type of parental behavior: Gottfried & Gottfried (1989); Openshaw,

continued

TABLE 8.1 Continued

Areas	Findings
Proposed Interactions	
8. Are there sex differences in the reactivity of older children to caregiver stimulation? (continued)	Thomas, & Rollins (1989); outcome measure: Cummings, Iannotti, & Zahn-Waxler (1985); Cummings, Vogel, Cummings, & El-Shiekh, (1989); or child IQ and ethnic group: Majoribanks (1981).
9. Are there interactions between classroom characteristics and individual personality and/or ability (aptitude by treatment interactions)?	*Yes:* Freebody & Tirre (1985); McGivern & Levin (1983). *Complex* higher order aptitude by treatment interactions, varying across outcomes or multiple individual or environmental characteristics: Corno, Mitman, & Hedges (1981); Cronbach & Snow (1977); Dunkin & Doenau (1980); Gilmore, Best, & Eakins (1980); McCord & Wakefield (1981); Wade (1981); Webb (1989).
10. Are temperamentally easy infants more responsive, cognitively, to positive features of the environment than difficult temperament infants?	*Yes:* MacPhee & Ramey (1981); Wachs & Gandour (1983). *No:* Maziade, Cote, Boutin, Bernier, & Thivierge (1987).

What are the reasons for the inconsistencies documented in Table 8.1?

1. *Statistical*: Wahlsten (1990) has persuasively demonstrated how standard statistical tests may be very inaccurate when used to test for interactions due to the fact that traditional procedures such as analysis of variance may have relatively little power when testing for nonadditive influences.

2. *Methodological:* McCall (1991) has argued that some interactions may appear only at the extremes of individual characteristics, but extreme groups are often underrepresented in most research designs.

3. *Covariance*: Existing interactions may be masked by covariance between organism and environment (McCall, 1991).

4. The operation of *higher order interactions*: Simple interactions between single organismic and environmental characteristics are moderated by multiple organismic characteristics (e.g., age of subjects, nature of risk status) or by the types of outcome measures under study (Sackett, 1991).

5. *Nonhomogeneous syndromes*: Studies mix together children having a variety of different etiologies and different mechanisms into a single group with the same label (e.g., hyperactivity, learning disabilities; Ackerman, Anhalt, Holcomb, & Dykman, 1986; Dykman, Ackerman, Holcomb, & Boudreau, 1983).

6. Overall *random findings*: Some overall random findings when considered in isolation could be viewed as examples of organism-environment interaction.

Evidence for Specific Organism-Environment Interactions

INTERACTIONS BETWEEN ENVIRONMENT AND
BIOLOGICAL RISK

Two major questions have driven research in this area. Each of these is considered in turn.

(1) Are children with a history of known or probable biological risk more adversely influenced by stress? Cutting across a diverse set of approaches, ranging from psychoanalytic theory (Benjamin, 1961; Greenacre, 1952) through modern-day behavior genetics (Kendler & Eaves, 1986), is the idea that, the stronger the genetic predisposition toward abnormal development, the less environmental stress is required to produce developmental deviations. A number of studies have supported this general hypothesis. For example, the negative relation between family stress and children's behavior problems or social competence is significantly exacerbated for children of depressed mothers (Hammen et al., 1987). Similarly results of a large-scale adoption study indicate that only 8% of the biological offspring of schizophrenic parents who were reared in seriously disturbed

adoptive families were rated as mentally healthy, whereas 63% were rated as severely disturbed; in contrast 23% of the offspring of nonschizophrenic biological mothers who were reared in severely disturbed adoptive families were rated as healthy, whereas only 37% were rated as disturbed (Tienari et al., 1985). Other studies have looked at the subsequent criminal behavior of children from criminal or noncriminal biological families who were adopted into families with or without a history of criminality. When only postnatal factors were operative (adoptive family criminality), offspring had 2 times greater risk of petty criminality than offspring with no risk; when only biological factors were operating (biological family criminality), the risk was 4 times as great. When both postnatal and biological factors were operative, however, the risk of offspring criminality was 14 times greater (Cloninger, Sigvardsson, Bohman, & VonKnorring, 1982). Within the same sample adopted offspring of criminal biological parents who had multiple temporary placements after birth were at higher risk for either subsequent criminality or alcohol abuse than multiadopted offspring from nonbiological criminal families (Cloninger et al., 1982). This pattern does not hold only for psychopathology. The degree to which adoption influences children's cognitive development may be moderated by the cognitive level of the biological parent. Thus Willerman (1979) has reported that 44% of adopted children whose biological parents had high IQs themselves had IQs greater than 120, compared with 0% of adopted children whose biological parents had low IQs; in contrast 15% of adopted children whose biological parents had low IQs had IQs below 95 compared with 0% of adopted children from high IQ biological parents.

While the available evidence suggests the operation of synergistic (multiplicative) interactions between biological predisposition and subsequent environmental stress, several possible exceptions need to be noted. Using an adoption study, with criminality of the biological and adoptive parents as the critical dimension, Mednick, Gabrielli, and Hutchings (1984) report that, while the overall direction of results suggested the possible existence of an interaction, the interaction term was nonsignificant when analyzed statistically. This discrepancy may be due to the possibility that the log-linear statistical procedure used by Mednick et al. may be less sensitive to interactions (Rutter & Pickles, 1991). A somewhat more complex set of findings has been reported by Walker et al. (1989). For adolescent boys, as would be predicted, the combination of parental schizophrenia plus parental maltreatment resulted in significantly greater rates of aggressive delinquency than either schizophrenia or maltreatment taken in isolation. In contrast, for girls, there were few if any significant findings, in terms of either main effects or interactions. As noted in Table 8.1 one possible explanation for these sex differences is that males may be more biologically vulnerable than females; hence interactions between predisposition and stress may be more likely to appear for males.

In general the above evidence does support the hypothesis that children who are biologically vulnerable due to parental psychopathology may be more sensitive to environmental stress than children without this type of biological predisposition. As noted in the previous chapter, however, environmental stress also tends to covary with parental psychopathology. Thus it is not clear whether the observed results are a function of at-risk children being more sensitive to environmental stress or whether children of disturbed parents also may be receiving more environmental stress. It could be argued that the adoption study design used in some of the above studies would get around this covariance problem. Even in adoption studies, however, evidence for selective placement exists, such that children from disturbed or low IQ families are more likely to be adopted into more disturbed (Tienari et al., 1985) or lower IQ (Willerman, 1979) families.

(2) Can positive environmental stimulation act to buffer (protect) children with a history of biomedical risk? Ample evidence supports the proposition that, even if a child is at definite risk due to biological deficits, positive features of the environment may act to buffer the child (Sameroff & Chandler, 1975). One of the most dramatic demonstrations of this buffering process is seen in a paper by Beckwith and Parmelee (1986), who reported that preterm infants showing certain electroencephalograph (EEG) patterns had poorer intellectual performance at 8 years of age than

preterm infants who did not have these risk patterns. The one exception was EEG risk preterms who were reared in homes where caregivers were characterized as highly responsive and sensitive; these infants showed later intellectual performance that was equivalent to that shown by non-EEG risk preterms. A similar pattern of results is seen in a study by Mednick, Brennan, and Kandel (1988), who reported that children born with minor physical abnormalities (which may be a marker for minor central nervous system deficits) were at higher risk for later criminal behaviors than were children born without these abnormalities. There was, however, little evidence of higher criminality rates for those children with high numbers of minor physical abnormalities who were raised in a stable family environment. Similar buffering patterns also have been shown for children at biological risk due to parental *schizophrenia* (Tienari et al., 1985), *alcoholism* (Cloninger, Bohman, & Sigvardsson, 1981), *prematurity* (Bradley, Caldwell, Rock, Casey, & Nelson, 1987; Drillien, 1964), or *inadequate nutrition* (Wachs et al., in press-a). Intervention studies also show a similar pattern of results. For example, Breitmayer and Ramey (1986) report that preschool children with low Apgar scores at birth averaged 12 IQ points lower than children with optimal Apgar scores. For those low Apgar children who received a program of psychosocial intervention, however, there were no IQ differences as a function of Apgar scores. Put another way, 44% of untreated low Apgar children were diagnosed as mentally retarded, as compared with only 14% of low Apgar children who received psychosocial intervention.

ENVIRONMENTAL HISTORY AS AN INDIVIDUAL
DIFFERENCES CHARACTERISTIC

While we traditionally tend to think of the organism part of organism-environment interaction as essentially based on intrinsic individual characteristics, it is also possible to think of the environmental history of the individual in a similar way. Within traditional psychoanalytic models, the child's earlier relations with his or her parents have been viewed as governing the way in which the child reacts when faced with later psychosocial stressors (Nunberg, 1955). Similarly, as suggested by Sroufe (1983), early relationships define expectancies the child has about other people and about the outcome of the child's behaviors. These early expectancies can function as "internal models," which become as much a part of the child's individual characteristics as does temperament or risk history. For example, children who are reared in more adequate environments may develop a stronger sense of mastery, self-confidence, and basic trust, which the child can then use when faced with subsequent stressors (Murphy & Moriarty, 1976).

If the above framework is correct we should be able to find evidence indicating that earlier positive experiences with the environment can act as a buffer, protecting children against later environmental stresses. A number of studies do support this hypothesis. Available evidence indicates that the risk of mental disorder for children living in disruptive homes or having parents with psychiatric disorders is markedly reduced if these children have a good relationship with at least one of their parents (Garmezy, 1987; Rutter, 1981a). Similarly a history of frequently shared activities between mother and child may buffer children against the negative effects of work stress when their mothers shift from part-time into full-time employment (Moorehouse, 1991). Werner and Smith (1982) have eloquently demonstrated how different experiences at different points in time relate to the child's ability to deal with later stress. Critical mediators that emerged in the Werner and Smith analysis included a lack of crowding and adequate caregiver attention in infancy, a warm positive relationship with a sibling in childhood, and family structure, cohesiveness, and social support in adolescence. Other studies have illustrated how securely attached children show more adaptive responding when faced with later environmental challenges than do insecurely attached children (Lewis, Feiring, McGuffog, & Jaskir, 1984; Sroufe & Egeland, 1991).

While available evidence supports the notion that positive early experiences have the potential to buffer the child against later environmental stresses, what is less clear is whether *early stress* can mediate the impact of later environmental influences. There are two possible hypotheses that have been noted here. The first is that early stress may *sensitize* the child, making the child more reactive to later stress situations. In contrast to sensitization there is also the possibility of early stress *"steeling"* the child, such that the child becomes less vulnerable to later stressors (Rutter, 1981a).

In general there is little evidence available that would enable us to distinguish between the sensitizing versus steeling hypotheses in childhood, though what evidence is available does tend to favor sensitization rather than steeling. For example, Anthony (1978) has hypothesized that the experiences that produce early insecurity may predispose the child to be resistant to later change, and to have higher levels of anxiety when faced with subsequent environmental changes. An excellent clinical example illustrating Anthony's hypothesis is seen in Pavenstedt's (1967) description of how preschool children from multiproblem disorganized families have difficulty varying their routines when placed in better environments. Pavenstedt suggests that this resistance to change is based on the continued operation of earlier coping mechanisms, which the child has used to provide his or her own consistency when the rearing environment has been totally chaotic. Similarly Finkelstein, Gallagher, and Farran (1980) have argued that exposure to excess noise in infancy may lead the child to screen out later auditory stimulation, including developmentally facilitative language stimulation. Grych and Fincham (1990) have hypothesized that children exposed to early conflict may be more likely to perceive conflict later in life when placed in emotionally ambiguous situations. Mentally retarded children with a history of adult rejection and abuse tend to be more sensitive to adult behavioral cues than retarded children without these histories (Zigler & Balla, 1982). In their study of children growing up under extreme social disadvantage, Werner and Smith (1982) report that more stress exposure was associated with a greater probability of deviant behavior when children were faced with later stresses.

In contrast, while there is ample evidence available from the infrahuman literature showing that mild stress in infancy may make animals more stress resistant later in life (see Thompson & Grusec, 1970, for a review), aside from case studies (e.g., Murphy

& Moriarty, 1976) there are few analogies to this phenomena reported in human developmental studies. One of the few examples that does exist is based on evidence indicating that the positive impact of encountering later challenges in new settings that recognize individual accomplishment is greater for individuals who, earlier in their lives, had experienced family economic stress (Bronfenbrenner, in press).

Not only is little evidence available on this question, but what evidence is available suggests that a dichotomy between early stress sensitizing or steeling children may be somewhat over-simplistic. In several papers Cummings and his colleagues (J. Cummings, Pellegrini, Notarius, & Cummings, 1989; Cummings, Zahn-Waxler, & Radke-Yarrow, 1981) have reported that pre-school children exposed to high levels of marital conflict were more likely to be reactive when faced with simulations of adult conflict than were children who did not have this history. The nature of the children's reaction varied considerably, however, with some children showing increased aggressive behavior or distress, whereas other children showed increases in expression of concern or prosocial behavior. A number of factors may influence what strategies develop in reaction to early stress. As suggested by Murphy and Moriarty (1976), moderate levels of stress are most likely to lead to steeling later in life. Too much stress may overwhelm the child, while too little stress may predispose the child to adopt a passive coping style, thus resulting in sensitization. The way the child copes with the stress also may be important. Rutter (1981c) has argued that, if children can successfully cope with early life stress, they may be steeled, whereas if children cannot cope with early life stress, they are more likely to be sensitized. The types of coping mechanisms the child uses to deal with early stress also may be important. If the child characteristically uses less mature coping mechanisms, this may increase the probability of later sensitization (Grych & Fincham, 1990). This may be particularly true if the negative coping mechanism the individual uses is one for which he or she has a biological predisposition, such as aggressive behavior (Olweus, 1986). What the above pattern suggests is that the question of sensitization or steeling may be best conceptualized as a form of higher order interaction, with multiple determinants governing whether and how the child is sensitized or steeled later in life.

SEX DIFFERENCES IN REACTIVITY

While the number of available studies looking at sex differences in response to environmental stimulation is relatively large, for the most part a confusing and inconsistent pattern of results emerges (see Table 8.1). There are three areas where there appears to be at least relative consistency. By *relative consistency* I mean that the preponderance of evidence suggests at least some degree of consistent sex differences in reactivity. Even within these islands of relative consistency, however, the heterogeneous mixture of predictor and outcome variables found for different studies makes it extremely difficult to develop an understanding of why sex differences in reactivity are occurring or what these differences tell us about process.

The first area of relative consistency involves crowding, with *crowding* being defined in a variety of ways such as sib spacing, rooms to people ratio, or experimental manipulations of density. Going from infancy (Wachs, 1979) to the preschool years (Koch, 1956a, 1956b, 1956c; Shapiro, 1974) through childhood and adolescence (Aiello, Nicosia, & Thompson, 1979), there is a consistent set of findings indicating that males appear to be more adversely influenced by crowding than females. Recent data suggest that this finding may hold cross-culturally as well (Wachs, Bishry, Sobhy, & McCabe, 1991). While the preponderance of evidence suggests a greater reactivity to crowding by males, there are some exceptions. Loo (1972) has reported a complex interaction, suggesting that manipulation of density conditions appears primarily to affect aggression in preschool males and dominance behavior in preschool females; Wachs (1987b) has reported no consistent sex differences in mastery as a function of crowding in the home. Further, while males and females do not appear to encounter different amounts of crowding, not all studies listed above actually tested to determine whether reported sex differences were statistically significant (e.g., Koch, 1956a, 1956b, 1956c; Shapiro, 1974). Given these ambiguities we can best conclude that the evidence for sex differences in reactivity to crowding is suggestive, though not definitive.

In terms of the second area, there are a number of studies indicating that male infants' development is more likely to be enhanced by a sense of control over their environment, with *control* being defined in terms of factors like maternal responsivity to the child (Martin, 1981; Rutter, 1983b) or in terms of the child's actual control over physical parameters of the environment (Gunnar, 1978; Gunnar & Donahue, 1980). This pattern does not occur for all outcomes, however (Martin, 1981). Further, there is also consistent evidence suggesting that male infants do better than females when adults *mediate* environmental stimulation (Wachs, 1984, 1987b; Yarrow, Goodwin, Mannheimer, & Milowe, 1973). At least on the surface adult mediation of the environment would seem to be contradictory to the need to have a sense of control over the environment. While it is possible to argue that male infants may do best if they can control when adults mediate the environment, currently there is no direct evidence to test this potential higher order interaction.

The final area where there is some consistency involves *parental loss*. What the evidence suggests is that, at least prior to adolescence, boys appear to be more adversely affected than girls by loss of a parent (usually the father), regardless of whether this loss is due to *divorce* (Bergman, 1981; Fry & Scher, 1984; Gurdinbeldi & Perry, 1985; Hetherington, Cox, & Cox, 1979, 1985; Rutter, 1981b), *death* (Levy-Shiff, 1982), or *separation* (Pedersen, Rubenstein, & Yarrow, 1979). In studies that do not show this pattern, it is often unclear whether another male has replaced the father in the home (e.g., Sciara, 1975). While reactivity to parental loss does appear to be moderated by sex of child, the evidence also suggests that sex differences in reactivity are only part of the story. Age of child also may be important. When divorce occurs during adolescence, available evidence suggests either that there appear to be no sex differences in reactivity (Forehand, McCombs, Long, Brody, & Fauber, 1988), or that sex differences in reactivity are moderated by the stage of the divorce process and the outcome measures used: Girls show internalizing problems primarily in response to marital distress prior to separation; boys show externalizing problems primarily after parental separation (Doherty & Needle, 1991). The nature of the family context following parental loss also may be relevant, with research indicating that preadolescent boys may benefit more than preadolescent girls from their mother's remarriage (Hetherington et al., 1985).

While at least part of the variability in reactivity to parental loss appears to be moderated by sex differences, this pattern of results tells us little about why these sex differences are occurring. Specifically, it is unclear whether the above relations reflect greater male vulnerability or differential treatment of males and females in single-parent families. For the most part few of the available studies have tested for the possibility of differential treatment, though both Pedersen et al. (1979) and Levy-Shiff (1982) do report that tests for differential treatment did not indicate that males and females were being treated differently. Again what we appear to have is a set of reasonably consistent findings that suggest the possibility, but only the possibility, of sex differences in reactivity.

INDIVIDUAL CHARACTERISTICS

Temperament. There are highly consistent findings supporting the hypothesis that infants or children having *difficult temperamental characteristics* may be more reactive to environmental stress than are infants with easy temperamental characteristics. This degree of consistency is particularly surprising, given the widely varying types of stress measures and measures of difficult temperament that have been used in different studies. Several studies (Langmeier & Matejcek, 1975; Wachs, 1987b; Wachs & Gandour, 1983) have reported that difficult temperament infants show greater reactivity to environmental noise and/or crowding than do infants characterized as having an easy temperament. Crockenberg has reported that the impact upon infant compliance or attachment of environmental stress factors, such as low social support (Crockenberg, 1981) or maternal anger (Crockenberg, 1987), is magnified if the infant has a difficult temperament; in contrast easy temperament infants appear to be more buffered against these types of stresses. Barron and Earls (1984) have reported that the relation of family stress to behavior problems in 3-year-old children is magnified if the child is characterized as being temperamentally "inflexible." Partial confirmation of this pattern comes from a study by Dunn, Kendrick, and MacNamee (1981), who report that, while there were no interactions involving problem behaviors or aggressive behaviors, the level of crying, withdrawal, or presence of sleeping problems associated with the birth of a younger sib was magnified if the older preschool age sib was highly intense or had predominantly negative moods. Looking at the reactions of preschool and school age children to stressful medical procedures, Lumley, Ables, Melamed, Pistone, and Johnson (1990) have reported that low adaptable children show significantly higher levels of distress when their mothers are emotionally unavailable than do more adaptable children. Finally, Hetherington (1989) reports that, even when family support was available, 10-year-old difficult temperament children showed less adaptive behavior with increasing family stress; in contrast, under these conditions, temperamentally easy children showed more adaptive behavior when exposed to moderate degrees of family stress. It could, of course, be argued that this set of findings reflects different environments provided to children with difficult versus easy temperaments. In a number of the studies cited (Crockenberg, 1987; Wachs, 1987b; Wachs & Gandour, 1983), however, this hypothesis was tested and was not confirmed, suggesting genuine differential reactivity rather than different environments.

There is also a surprising amount of consistency in studies investigating the potential mediating effects of *activity level* upon reactivity to the environment. The initial evidence on this question came from a clinical study by Escalona (1968), who did detailed cross-sectional observations on a sample of high and low active infants. Escalona's data suggested that high active babies may be more reactive to lower levels of stimulation, whereas less active babies may need stronger amounts of stimulation to facilitate their development. These results, while provocative, were problematic due to the small sample size and the fact that the activity and behavioral ratings were not always independent. Initial support for Escalona's observations came in a study by Schaffer (1966) reporting on infants' reactions to short-term hospitalization. If one assumes that hospitals are not highly stimulating places, Schaffer's data, indicating that highly active infants are more resistant to the negative consequences of short-term hospitalization than low active infants, could be considered as support for the conclusion that active babies need less stimulation. Again, however, the possibility of differential treatment was not ruled out. Differential treatment was ruled out in two subsequent studies, which indicated that high active infants showed higher levels of mastery motivation (Wachs, 1987b) and exploratory play competence (Gandour, 1989) when their parents were less stimulating or less emotionally involved; in contrast low active infants showed higher mastery or more exploratory play competence when their parents were highly stimulating or highly involved. In neither study was there any difference in the environments encountered by high and low active infants. The Gandour study is particularly noteworthy also in terms of indicating that the predicted interaction between temperament and environmental stimulation occurred only for activity and did not occur when the sample was divided on the basis of either sex or persistence. Again, given differences between studies on both the predictor and the outcome variables, this degree of consistency is quite remarkable.

Although less evidence is available, there are some suggestions in the literature that temperamental differences in *reactivity* to stimulus threshold or *stimulus sensitivity* also may lead to differential reactivity to the environment. Based on clinical observations Murphy and Moriarty (1976) suggested that children with moderate levels of reactivity typically show greater learning when faced with environmental demands, whereas children with high levels of reactivity tend to show behavioral disorganization. Strelau (1983) has reported evidence from a number of Eastern European studies indicating that high sensitivity children tend to be disorganized by environmental stress, whereas low sensitivity children tend to be energized. For example, poor school grades further disrupted the school performance of children who were characterized as high stimulation sensitive; in contrast poor school grades tended to enhance subsequent performance of children who were characterized as low in stimulus sensitivity.

While the above findings generally support the existence of temperament × environment interactions, particularly those involving difficult temperament and activity level, it is important also to note the results of a study by Plomin and Daniels (1984). These authors analyzed for temperament × environment interactions in both the infancy and the preschool periods across a wide range of environmental and temperamental variables. Basically they report few if any significant interactions. These nonsignificant findings may be attributable to methodological factors (e.g., the environmental measures used by Plomin and Daniels consisted of both parent questionnaires and direct observations, whereas the overwhelming majority of studies showing significant interactions used direct observation). I would argue, however, that these nonsignificant results illustrate a more fundamental point. As noted earlier, interactions appearing in both the infrahuman and the human biobehavioral literature typically involve relatively specific combinations of organismic and environmental characteristics. Similarly the temperament-environment interactions reported above involve relatively well-defined temperamental dimensions (difficultness, activity level) and environmental characteristics (specific indices of stress or intensity of stimulation). Only one aspect of temperament (activity), and neither of these environmental dimensions, was assessed in the Plomin and Daniels study. To the extent that interactions involve *specific individual and environmental characteristics,* it should not be surprising that studies that do not measure these characteristics are less likely to report significant interactions. This emphasizes the importance of designing research that is targeted to look for specific interactions rather than looking for interactions per se. Based upon the available evidence the combinations of difficultness/stress and activity/intensity of stimulation should be prime areas of focus in future research on temperament-environment interaction (with perhaps further focus on the question of potential interactions between sensitivity and environmental demands).

Children with externalizing behavioral characteristics. Although available evidence is limited, there are some suggestions that defiant or aggressive children may react differently to the environment than children without these characteristics (Rosenthal & Lani, 1981). For example, Parpal and Maccoby (1985) report that the behavior of preschool children varies as a joint function of the nature of the play situation the child is involved in as well as the child's characteristic level of compliance with adult requests. Dodge (Dodge et al., 1980) has reported evidence indicating that aggressive children are more likely to interpret ambiguous social stimuli in an aggressive manner than are children without a history of aggressive behavior. Similarly aggressive boys also have a higher probability of reacting aggressively to situations they interpret as aggressive than do nonaggressive boys (Dodge, Coie, Pettit, & Price, 1990). These patterns are particularly likely to occur if boys are both hyperactive and aggressive (Milich & Dodge, 1984). Dodge's results suggest that specific patterns of stimuli may have different meanings for children, depending upon their characteristic level of aggression. While this pattern may be partially dependent upon the types of stimuli used (the pattern is more likely to appear in open-ended situations than in forced choice situations; Milich & Dodge, 1984), similar patterns also have been noted for aggressive girls (Dodge, 1986) and in a cross-cultural test in Japan (Tachibana & Hasegawa, 1986).

Preference for stimulation. There are several studies that have looked at the individual's own preferences for different types of environmental input, as these enter into interactions. For example, Carbo (1983) has reviewed data indicating that students who prefer quiet do better in quiet classes, whereas students who prefer sound do better in classes where background talking is allowed. Students who prefer dim light read better in dim light whereas students who prefer bright light read better in bright light. Schaffer and Emerson (1964) have reported that infants characterized by their mothers as "cuddlers" show greater positive reactivity to physical contact stimulation than infants characterized as noncuddlers. Children's likes and dislikes also may act as individual difference mediators. For example, Slife and Rychlak (1983) have reported that children show greater modeling when they are exposed to liked materials or liked actions than when they are exposed to disliked materials or actions.

While the above evidence could suggest that individual preferences can act as a mediator of environmental influences, there is a tricky conceptual point that hinders easy acceptance of this interpretation. Specifically it could be argued that the above findings may represent an example of *active covariance* (niche picking). For example, Parrinello and Ruff (1988) report that infants who are low in attention to novel objects do better when adults actively mediate the environment for these infants, whereas adult mediation is less important for infants who are high in attention to novel objects. This could suggest the possibility of interaction; however, Parinello and Ruff also note that infants who are low in attention to novel objects tend to focus much more on the adults in their environment than on objects, suggesting the possibility of self-imposed differential exposure.

Conclusions

Given the above evidence, what conclusions can we draw about the existence of interactions between specific individual characteristics and specific environmental influences? Based on the evidence reviewed there appears to be a consistent body of evidence illustrating how positive features of the environment can act to buffer the individual child against exposure to biological or subsequent environmental risks. Similarly, we also have a consistent body of evidence illustrating how difficult temperament or biologically vulnerable children appear to be more reactive to environmental stress than are temperamentally easy or nonvulnerable children. While less evidence is available, a number of studies also have shown how less active children appear to need more intense stimulation or caregiver involvement than do more active children.

On the other hand there are a large number of studies yielding totally inconsistent, specific organism-environment interactions. For example, it simply is impossible to say whether biologically at-risk children need similar or different types of stimulation than nonrisk children or whether stress has a greater impact on males or females.

While there are a number of possibilities for why the findings are so inconsistent, I would argue that a major reason may be conceptual. Interactions may in fact exist, but they may not be the simple lower level organism × environment interactions we have been reviewing. Rather than lower level interactions, what may in fact be operating are *higher order interactions,* encompassing multiple individual and contextual parameters (Cronbach & Snow, 1977; Sackett, 1991; Wachs, 1991d). Using sex differences as a marker, we can illustrate this point by reconsideration of the question of whether males are more vulnerable to parental loss than are females. While evidence reviewed earlier suggests that males are more vulnerable than females to parental loss, available evidence also indicates that the degree of vulnerability may be moderated by a variety of other factors, including age of child and nature of the family context after loss. Thus what initially appears to be a simple lower order sex × loss interaction may actually reflect the operation of a higher order, multidimensional organism × environment interaction. Similarly, it could be argued that sex differences in reactivity to stress reflect not so much differences in *vulnerability* to the environment as differences in *reactivity* to the environment. Thus males and females may be equally vulnerable to environmental stress but may show it in very different ways (Doherty & Needle, 1991; Zaslow & Hayes, 1986). Specifically, evidence suggests that males may be more likely to react to stress through use of externalizing symptoms such as anger, whereas females appear to be more likely to react using internalizing symptoms such as anxiety (Emery, 1982; Grych & Fincham, 1990;

Lewis et al., 1984). This suggests the existence of an organism × environment × outcome higher order interaction.

To the extent that higher order interactions exist, this suggests the need for major reconceptualization, not only in terms of how we view interactions but also in terms of how we design studies to test for these interactions. Traditionally we have viewed interactions as operating for all members of a specific population (e.g., all males will be more vulnerable). To the extent that higher order interactions are operating, however, interaction processes may be relevant only for a specific subsample of a given population (e.g., greater vulnerability will be shown for preadolescent males living in single-mother households) or for a given outcome (e.g., externalizing behaviors for males). Under these circumstances Rutter (personal communication) has suggested that behavioral researchers interested in the study of interaction should avoid looking for generalized differential reactivity. Rather researchers should focus on specific sets of individual characteristics, specific sets of environmental variables, and a hypothesized process mechanism that targets specific outcomes as being most sensitive to the interactions under study. Potential sets of individual characteristics have been delineated all through this chapter; potential sets of environmental characteristics may be derived from material presented in Chapters 3 and 4. Theory-driven process mechanisms are harder to come by (see some discussion of this point in Chapter 9). Two potential possibilities include differential vulnerability and differential variation in the child's ability to make use of environmental supports (Wachs, 1991d). One major advantage to approaching the study of organism-environment interaction in this way is that it then becomes possible to tailor the analytic procedure used to the specific process mechanism that is presumed to be operating, rather than using less precise global tests of interaction (for further discussion of this point, see Rutter, 1983b; Rutter & Pickles, 1991).

The importance of looking for specific process mechanisms goes beyond just the study of organism-environment interaction. What looks like differential reactivity on the surface may reflect the operation of a different process or even the simultaneous operation of multiple processes. One interesting example can be seen for the area of sex differences in reactivity. Given the possibility that males may have a greater probability of low level central nervous system damage (Bolter & Long, 1985; Zaslow & Hayes, 1986), males may be more likely to show problem behaviors than females. If preexisting differences between males and females are not measured, and we assess male female differences *after* exposure to factors like stress, day care, or divorce, the results may suggest differential reactivity. Rather than differential reactivity, what we really have demonstrated is stability of preexisting differences in behavior that have little to do with stress, divorce, or day care (Gamble & Zigler, 1986). An excellent illustration of this problem is seen in a longitudinal study by Block, Block, and Gjerde (1986), who did an intensive evaluation of 3-year-old children and then followed the children up through age 14. Over this time span a number of divorces occurred in their sample. If we simply look at the children's development after the divorce, the results could be said to fit the expected pattern of boys showing more personality problems than girls. If we look at the sample at age 3, however, the results show that boys tended to have more personality problems than girls, even up to *11 years prior* to the divorce. It is possible that factors like stress or day care may amplify existing sex differences in behavior (Block et al., 1986) but, for most studies, we cannot rule out the possibility that group differences may precede and not follow specific environmental events.

Other examples of alternative process also exist. For example, a clear interpretation of findings indicating that males may be more vulnerable than females to day care or parental loss is difficult, in good part because all too few studies analyze whether there is different treatment of males and females in day-care situations or in single-parent families. Certainly sex differences in treatment of males and females are not unlikely, given that these differences appear even in subcultures that stress equality of treatment (Weisner & Wilson-Mitchell, 1990). Similarly, evidence cited earlier suggests a potentially greater vulnerability to environmental stress for children whose parents have a history of mental illness or conduct disorder. Does this finding reflect a synergistic interaction between biological and environmental risk, or does it reflect the naturally occurring covariance between biological and environmental risk (see Chapter 7), resulting in these children being exposed to more risk factors? What this means is that finding individual differences in reactivity to specific environmental contexts represents *only a first step.* The next critical step is to determine whether the observed results reflect genuine differential reactivity to similar environments (organism-environment interaction), initial group differences that are maintained across time, differential treatment of different individuals within a given context (reactive organism-environment covariance), or some combination of all these processes (Zaslow & Hayes, 1986).

Given all of the above, there are a number of specific conclusions we can draw about the process of organism-environment interaction and about the nature of environmental influences upon development.

(1) There are too many examples of both nonspecific and specific organism-environment interactions simply to dismiss evidence for interaction as reflecting nothing more than random error variance. What this means is that both our theories and our studies of environment must be broadened beyond simple main effect models to take account of the mediation of environmental influences by nonenvironmental factors as well as by earlier environmental influences. This also means that we need to pay special attention to the unique methodological and statistical problems involved in the study of organism-environment interaction.

(2) We need not assume that all individual characteristics or all aspects of the environment are sensitive to interactions. Rather than looking for interactions in a global manner, we should be targeting specific combinations of organismic and environmental characteristics. These combinations should be derived either from theory or from previous research.

(3) We need to consider seriously the possibility that interactions may be higher order rather than lower order in nature. If this assumption

is correct, then this is all the more reason to move away from simple main effects models of environmental action. If higher order interactions exist, then the operation of these interactions is most likely to be revealed when research is theory driven and when the theory that drives the research is one that allows for the possibility of higher order interactions (Wachs & Plomin, 1991).

(4) Up to the current time we have been treating organism-environment covariance and organism-environment interaction in isolation or even as antagonistic. In reality what is probably happening is that both are co-occurring simultaneously. What this means in practice is that variability in developmental outcomes may need to be viewed as a function of the simultaneous operation of both covariance and interaction. One of the few existing examples illustrating this simultaneous co-occurrence is seen in a paper by Rahmanifar (Rahmanifar et al., 1992) on nutritional, biomedical, and environmental predictors of infant alertness (see Figure 8.1).

What is illustrated in Figure 8.1 is that, while there is a main effect relation between maternal nutrition and infant alertness, it would be a mistake to interpret variability in alertness as due just to variability in nutrition. Also shown in this figure are synergistic interactions between maternal food intake and measures of crowding in the family and infant morbidity, both of which substantially increase predictive variance. In addition unique predictive variance also is seen for the covariance between maternal food intake and maternal vocal stimulation of the infant.

The implication of this final point is consistent with what was stated earlier. To the extent that there is simultaneous co-occurrence of organism-environment covariance and interaction, main effect models of environmental action are most likely to be oversimplistic and not reflect the actual processes through which environmental variability is translated into variability in development.

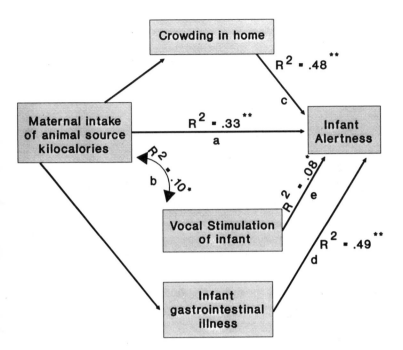

Figure 8.1. Variables Influencing Amount of Time Infant Is in an Active Alert State

NOTE: Path *a* refers to the magnitude of the direct relation between maternal food intake and infant alertness. Path *b* refers to the covariance between maternal food intake and the vocal stimulation of the infant. Paths *c* and *d* refer to the magnitude of the combined influence of maternal food intake and either home crowding or infant morbidity upon alertness. Path *e* refers to the magnitude of the unique relation between maternal vocal stimulation and its covariate (maternal food intake) upon alertness.

The Nature of Nurture

Implications and Applications

In terms of our understanding of both the struc-ture of the environment and the processes under-lying environmental influences upon behavior and development, we appear to have come a long way from Phase I studies on "the environment." A summary of the major points contained in this volume will illustrate not only how far we have come but also where we need to go.

The Nature of Nurture

To understand the role of environmental influences, it is critical to understand that the environment operates as part of a more general system (for purposes of this chapter I am defining a *system* as a set of variables that have the potential to influence develop-ment, and that are organized within a conceptually meaningful framework; Wachs, 1991d). This more general system encom-passes hierarchically organized multiple environmental compo-nents, nonenvironmental individual characteristics, and biosocial-biobehavioral parameters. While we traditionally refer to the *environment,* a more appropriate label to use when discussing the nature of nurture would be the *environmental system.*

Structurally the environmental system encompasses *multiple levels.* Higher levels of the environmental system can influence both the nature and the impact of the environmental system at lower levels. The direction of influences within the environmental system, however, is not necessarily from higher to lower levels. The environmental system is potentially bidirectional, both within and across levels.

The environmental system functions not only in space but also in *time.* Thus there appear to be age periods when the individual is more sensitive to *specific* environmental influences than at earlier or later times. This may be particularly true for those developmental outcomes that can be described as *"universals"* (Horowitz, 1987), or that are governed by experiences that individuals typically encounter at a given point in development (*experience expectant stimulation;* Greenough et al., 1987). While early experiences encountered by the individual may have long-term effects, particularly if these experiences serve as a filter for later experiences, the impact of the environmental system is not restricted just to the role of early experience per se. Depending upon the outcome variable under study, concurrent or cumulative aspects of the environmental system may be more important than earlier experiences.

The environmental system also encompasses the idea that different aspects of the environment influence different aspects of development (*environmental specificity*). The operation of environmental specificity may be rooted in the interface between biology and development, particularly when we look at the influence of unique sources of environmental input upon localized areas of the central nervous system (*experience dependent stimulation;* Greenough et al., 1987).

It also seems clear that the environmental system does not operate in isolation. Various dimensions of the environmental system can *covary* with a variety of nonenvironmental factors (e.g., nutrition, difficult temperament), all of which also have the potential to influence development. Covariance means that children in certain environments (e.g., poverty) have a greater than average probability of encountering specific nonenvironmental covariates, such as inadequate nutrition, or that children with certain characteristics (e.g., difficult temperament) may be more likely to elicit certain kinds of reactions, such as parental hostility. The existence of organism-environment covariance does not mean that the impact of the environmental system upon development can be reduced to the impact of the covariate. The very nature of the environmental system rules out such reductionism. What the existence of covariance does mean is that, the greater the degree of covariance among risk factors, the more likely the child is to be at risk. On the other hand, if child characteristics covary with caregiver preferences (fit), the child may be more likely to be buffered against risks.

In addition to covariance the environmental system also encompasses the *interaction* of environment with individual characteristics or biological factors, such that different individuals may react differently to objectively similar environmental stimulation. For the most part the study of organism-environment interactions has involved only one characteristic of the organism and one dimension of the environment (lower order interactions; Wachs & Plomin, 1991). This approach may be oversimplistic, however. Just as higher level environmental variables may moderate the impact of lower level environment-development relations, higher order interactions also may serve to moderate the impact of lower order interactions.

Given the possibility that a picture equals 1,000 words, to save further space I have summarized my view of the environmental system in Figure 9.1. The system pictured has been simplified at specific points, though all of the main features of the environmental system are contained in this figure. With the description of the environmental system contained in the text and Figure 9.1, we can now turn to the question of the implications of this system for future research, theory, and application.

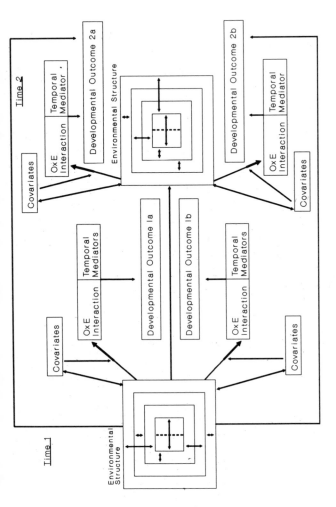

Figure 9.1. The Environmental System

NOTE: Double headed arrows refer to bidirectional paths of mutual influence. Single headed arrows refer to unidirectional paths of influence. O × E refers to organism-environment interaction influences. For simplicity's sake I have omitted paths of influence from the multiple levels of the environmental structure to developmental outcomes and the potential stability of covariates or developmental outcomes across time.

143

The Environmental System and Future
Research on Environmental Influences

A variety of suggestions about research methodology in specific domains has already been made in previous chapters of this volume. This chapter deals less with methodological aspects associated with particular parts of the environmental system and more with the research implications involved when we are looking at the environmental system per se.

In general most studies of environmental influences have looked at the environment in isolation, using Phase I or Phase II research strategies. In a historical sense this approach has been validated by our finding significant and even replicable patterns of environment-development relations (see Chapter 3). It is important to keep in mind, however, the distinction between statistical significance and effect size. It is entirely possible to find statistically significantly main effects and yet account for very little predictive variance. If we study the environment in isolation, and come up with significant environmental main effects that account for a substantial amount of variability in our outcome measure (greater than the amount of nonpredicted residual variance), an argument can be made for the validity of the environment in isolation approach. For those situations where we have statistically significant environmental main effects, but the amount of variance accounted for by these environmental main effects is relatively small, it may be critical to consider the possibility of moderator or boundary conditions, wherein strong environmental influences may occur only for certain subgroups of individuals or only within certain contexts (see Chapters 4 and 8).

For example, there are a number of studies looking at the relation of family interaction patterns to children's behavior disorders, which indicate that significant relations appear only when samples are divided on the basis of the type of psychopathology involved (Jacob, 1975) or by sex of the child (Block et al., 1988). One way this can happen is when we have a *disordinal interaction*, such that environment development relations are significant for two groups but the directionality is opposite (e.g., + for males,

– for females); when we combine the groups, the opposed pattern of relations cancels each other out. This situation also can happen when environment functions as a "catalyst," influencing outcomes only in the presence of other environmental events such as stress (e.g., Rutter, 1983b). Thus finding either nonsignificant or significant but small environmental influences should not necessarily lead us to the conclusion that environment is unimportant; rather the appropriate conclusion may be that an alternative Phase III research strategy needs to be used, namely studying the environment as a system.

While it is possible to determine by inspection of our data whether we need to focus on the environmental system, this knowledge will be of very little practical use if the researcher has not already collected sufficient information to test salient aspects of the environmental system. This leads to a second conclusion. Let us assume *at the outset* that we are dealing with an environmental system. Making this assumption demands that the researcher collect sufficient "side" information to illustrate system processes as part of the normal research process (Cronbach, 1991).

If we start out by assuming that we are dealing with an environmental system, two questions immediately follow: First, *what additional information should be collected?* Second, *how should this information be used?* In terms of collecting appropriate information, Bronfenbrenner (in press) has argued that studies of environmental influences should include at least two distinct macrosystems as part of the research design. I would go beyond Bronfenbrenner to argue that it would be to our advantage for Phase III environmental studies also to include sufficient information to test environment-development relations for two subgroups of individuals, and to include information on at least one potential covariate and one alternative context. The choice of which macrosystems, which individual characteristics, which covariates, and which contexts will depend upon the goals of the research. If the researcher wishes to use additional information only as a cross-check on the generalizability of observed environment-development relations, the focus should be on easily obtained and relatively less precise measures. In this situation the basic design could involve looking at whether environment development relations vary across caregiver educational level or social class (macrosystem proxies; Bronfenbrenner, personal communication) or are different for males and females (individual characteristics), are related to height (covariate proxy for nutritional history in disadvantaged populations), or parent report of medical history (covariate proxy for biomedical status in more advantaged populations), or vary as a function of number of hours spent in day care, or nursery school or as a degree of family stress or social support (context proxies).

Phase III studies that are designed to test specific hypotheses about the nature of the environmental system will require more precise and well-defined measures of the *macrosystem* (e.g., specific cultural groups), *individual characteristics* (e.g., temperament, studying children at different ages), *covariates* (nutritional intake, selected parent characteristics that may covary with the child characteristics under study), and *contexts* (quality of day-care situations). Obviously intermediate research strategies are also possible. For example, for researchers who wish to test specific hypotheses about a portion of the environmental system without necessarily testing the complete environmental system, the appropriate strategy would be to obtain less detailed information on those parts of the environmental system that are not the major area of focus along with more detailed measures for that part of the environmental system that is of primary concern.

Once the appropriate information is gathered, the next question is how best to use this information. Traditionally such information is subjected to statistical control, partialing out the role of factors like social class or parental educational level when studying environment development-relations. This use of statistical controls, however, is based on the assumption that environmental processes operate in the same fashion across different levels. This assumption has been questioned by a number of researchers (Bronfenbrenner, 1989; Cloninger et al., 1981). In addition, to the extent that obtained environmental measures may covary with individual or contextual factors, by use of statistical controls we may be partialing out the very environmental effects we wish to study.

Rather than routine use of statistical controls, a more appropriate approach would involve tailoring analyses to the specific questions that are being asked about the operation of the environmental system. For example, if we are interested in generalizability of environment-development relations, the appropriate analysis would involve comparison of these relations across different contexts or populations. An emphasis on covariance might involve looking at outcome as a function of the number of buffers or risks encountered by the child. Looking for higher order interactions, or for the simultaneous operation of covariance and interaction, may require the use of special tests for multiplicative, nonlinear functions (see Baron & Kenny, 1986; Roberts, 1986, for procedures for nonlinear analysis).

Regardless of the specific tests that are used, when looking at the environmental system one major problem that must be considered is the question of *statistical power* (the probability that our analysis will detect existing relations). In situations where there is low statistical power, there is always the possibility that critical contextual or individual moderators of environment-development relations may not be identified. Studies on the environmental system may be particularly vulnerable to low power, given that power decreases as the number of variables involved increases (Cohen, 1988). If the researcher does not substantially increase the sample size to match the increase in number of variables needed to adequately investigate the environmental system (*in addition to* the microsystem dimensions under study, add in two macrosystems, two individual characteristics, one covariate, and two contexts), the resulting reduction in power may lead to the incorrect conclusion that there are no significant macrosystem, individual, covariate, or contextual effects operating. If sample sizes are increased, then it may become difficult to use highly precise environmental measures, such as those based on repeated observations (Wachs, 1991a). This, of course, is a major disadvantage when the goal of the research is to study environmental influences. Fortunately there are some ways through which researchers can raise power while still using relatively precise measures of the environment.

1. At the level of research design, one obvious approach to maximizing power is through the use of *aggregated measures*. Aggregated measures are those that are based on multiple measurements obtained across occasions, observers, or related measures. Statistical aggregation reduces measurement error, which in turn increases power through reducing variability in sources that are unrelated to the phenomena under study (Cohen, 1988; Sutcliffe, 1980; Zimmerman & Williams, 1986). Thus the use of aggregated data may help to compensate for the potentially low power that can result when simultaneously studying multiple dimensions of the environmental system. Aggregation within a study is costly, however, in terms of increased time demands (e.g., multiple observations).

2. Aggregation techniques also can be used *across studies* (Wachs, 1991a). Specifically, if different research projects use a common set of environmental and outcome marker variables, it then becomes possible to aggregate across different studies, which has the effect of substantially increasing the effective sample size. Fundamentally, cross-study aggregation allows researchers to combine the preciseness found in individual small sample studies with the statistical power found in large sample studies. For example, by using a common set of measures of the microenvironment, and by aggregating these measures across six separate studies, Bradley et al. (1989, 1990) were able to test environment-development relations simultaneously across both different macrosystems (social class) and individual child characteristics (race, sex).

3. Another approach to increasing power would be to reduce the number of variables to be tested, as seen in the use of clustering procedures such as principal components analysis or scaling. Use of clustering procedures allows us to nest different levels of the environment within a single cluster (e.g., Block & Block, 1981; Steinberg et al., 1991). Clustering approaches may be particularly useful when various levels of the environment covary. Particularly for older children or adolescents, a complementary strategy would be to use individuals' perceptions of themselves or their environments as a means of clustering individuals or situations (Block & Block, 1981; Mischel, 1973).

4. Power also may be increased when we test specific, well-defined hypotheses, as opposed to more global tests of the influence of the "environment" (Maxwell, 1990). An additional advantage of generating a specific set of hypotheses for testing is that it then becomes possible to tailor the statistics used to the question under study (Rutter, 1983b; Rutter & Pickles, 1991). The importance of being able to test specific hypotheses emphasizes the need for *theory-driven research* on the environmental system, as a means of generating these specific hypotheses. In the absence of highly precise environmental action theories (see below), descriptive models or even descriptive research may be a good starting point to generate specific hypotheses that can be tested (Cronbach, 1991).

5. Finally, it may be important to recognize the possibility that use of the traditional *.05* level of statistical significance, as the criteria for defining the "truth" about a specific hypothesis, may be inappropriate when testing predictions about the environmental system. By emphasizing the need to avoid type I errors (accepting an invalid hypothesis as correct), we have all too often ignored the converse type II error (falsely rejecting a correct hypothesis). The historical basis for this statistical strategy has been traced by Bronfenbrenner (1991), who has argued for a radical solution, namely, looking for consistent patterns of results across studies regardless of the significance of the results of a single study. A somewhat more conservative approach (though in the same spirit) has been suggested by both Cronbach (1991) and Rutter and Pickles (1991). These authors recommend testing models of environmental action using *confidence intervals*. Use of confidence interval procedures allows us to retain potentially useful environmental models for further study, even though there may not be sufficient power within a single study to confirm the statistical "truth" ($p < .05$) of a specific model.

Implications for Theory

What should be clear from this volume is that main effect Phase II environmental theories, while appropriate for understanding the role of environmental influences within a specific context or for a specific population, are incomplete when applied to the question of understanding the nature of environmental influences per se. While it is possible to continue to develop main effect environmental theories, in the hope that someday we may be able to combine separate main effect theories into a more comprehensive theory, this approach is based on the assumption that the whole equals the sum of the parts. To the extent that the environment functions as a system, this assumption may not be correct. Systems appear to have unique organizational properties that operate independently of the specific parts of the system, so that the whole is likely to be much more than the sum of its parts (Sameroff, 1983). Thus it may be more advantageous to use existing descriptions of the environmental system as a starting point for developing more comprehensive Phase III theories of environmental action. What I have attempted to do in this volume is to develop some of the criteria for environmental systems that should be incorporated into future theories. These criteria include the following:

1. The environment should be seen both as a set of *distinct levels* as well as a set of *subunits* within each level. Theories should provide for bidirectional influences both across and within levels, such that environmental influences at one level can influence the nature of the environment and the nature of environmental action at other levels.

2. Theories should include a *longitudinal component,* allowing for the transmission of the environment across time, the moderation of environment-development relations through temporal factors such as maturation level of the organism, and specification of processes through which earlier experiences either can cumulate or can act to moderate ongoing relations between environment and development.

3. Theories should be *probabilistic* rather than deterministic in nature, in the sense that the relations between environment and development should never be viewed as fixed (Gottlieb, 1991). Rather theories should be designed to allow environment-development relations to be moderated by the developmental outcomes that are under study (specificity), by nonenvironmental

factors that covary with the environment, and by individual characteristics that can interact with the environment.

4. Environmental theories not only should enable us to describe the structure of the environment but also should encompass *process*, in the sense of allowing us to formulate hypotheses as to how variability in environment is translated into variability in development.

Do we have any existing environmental theories that meet these criteria? While some environmental theories meet some of the criteria, none meets all of the criteria. For example, the systems model described by Ramey, MacPhee, and Yeates (1982) has both a longitudinal component and a structural component, encompassing multiple levels of the environment. This model, however, does not clearly allow for bidirectional influences within levels, specificity, or the possibility that differential outcomes may result from the interaction between child characteristics and environment. The structural behavioral theory of Horowitz (1987) contains a longitudinal component, takes into account the interaction between individual child characteristics and environment, and allows for the possibility of specificity. The structural behavioral theory, however, is limited by its reliance on a bipolar concept of the environment (facilitative versus nonfacilitative). The ecological theory of Bronfenbrenner (in press) offers a detailed description of the multilevel nature of the environment and of the linkages between different levels, has a longitudinal component, and allows individual characteristics to moderate environmental action. Ecological systems theory, however, does not clearly deal either with the probability of specificity or with nonenvironmental covariance. The resilience theory of Werner and Smith (1982) does encompass multiple levels of the environment, individual mediators, a longitudinal component, and the possibility of covariance. Resilience theory, however, does not clearly allow for the possibility of specificity or the influence of macrosystem mediators upon lower level environmental processes.

Given that it is based on the research reviewed in this volume, it is perhaps not surprising that the environmental systems framework illustrated in Figure 9.1 may come closest to meeting all the criteria defined above. The environmental systems framework allows for temporal influences, specificity, covariance, and inter-action. It does not tell us, however, which individual or biological parameters will covary or interact, which outcomes will be influenced by which predictors, or how variability in individual characteristics or age translates into variability in development (process). In this regard the theories developed by Bronfenbrenner, Horowitz, or Werner and Smith are much more detailed in terms of delineating the operation of specific process variables.

The fact that no current theory of environmental action meets all of the criteria specified above is not necessarily a reason for despair. Indeed the fact that we have gone beyond main effects theories of the environment is in and of itself a cause for optimism, though not necessarily a reason for complacency. In terms of future progress, one way of improving existing environmental theories is to emphasize Phase III research that provides more detail about assumptions inherent in these theories. For example, given that Phase III environmental theories emphasize the importance of within- and across-level mediation of specific environmental influences, it is critical to establish the degree to which such mediation is actually occurring. Given that Phase III theories should allow for individual characteristics either to increase vulnerability or to buffer the child against stress, it is important to identify which individual characteristics can act as buffers, what aspects of the current environment are most likely to set off previously developed vulnerability, or how prior experiences can function either as buffers or as sources of vulnerability. Given that the nature of environmental systems allows for the operation of higher order interactions (e.g., sex by developmental stage; sex by stress by outcome), it is imperative to identify when higher order interactions are operating and what is the impact of these higher order interactions upon lower order environmental processes or organism-environment interactions. Further, given that organism-environment interactions may occur as a result of differential vulnerability, differential use of environmental opportunities, or differences in response patterns (Wachs, 1991d), tests of when and how these different processes are operative are critical, if we are to develop theories that do more than talk about interaction in very general terms. Finally, given that in many cultures the primary caregiver is not the child's parents, it may be important to expand existing theories by asking Phase III context questions, such as whether the same environment-development processes occur if sibs, grandparents, or nonfamily caregivers have primary responsibility for child rearing (Werner & Smith, 1982).

Applications

Over the past 40 years there has been an increasing emphasis upon manipulation of the environment as a means of enhancing various aspects of children's development. While there have been a variety of successes, it is perhaps fair to state that the overall outcomes have not been as positive as proponents of environmental influences would have hoped for; nor have they been as negative as predeterminists would have liked. I have argued (Wachs, 1990) that one of the major reasons for this inconclusive state of affairs is our assumption that the environment operates in a main effect sense, so that we assume that developmental problems will be remediated if we simply place the child in an "optimal" or an "enriched" environment. In fact there may be no such thing as an *overall* optimal environment (Benninger, 1978). Within a Phase I or Phase II framework it was logical to assume that interventions that succeeded with one population or within one context would work equally well across all populations or contexts. The fact that a set of environmental variables reliably facilitates improvements in specific developmental outcomes, however, does not necessarily mean that this set of variables will be equally facilitative under all circumstances. What is optimal may well vary as a function of age of child, individual child characteristics, outcome variables under consideration, characteristics of other aspects of the environment, or influence of nonenvironmental factors that covary with environment. Given what we now know about the Phase III nature of the environmental system, the time would seem ripe to begin to rethink our intervention strategies so that they accurately reflect the operation of the environmental system. There are a number of ways this might be done.

1. Given that the environment of the child consists of more than just the family or the classroom, it may be important to intervene simultaneously at multiple levels of the environmental system. The most obvious examples are seen in programs that deal simultaneously with the child and the family (e.g., Seitz, Rosenbaum, & Apfel, 1985). Do greater gains appear with multiple-level interventions? What little evidence is available on this question suggests that multiple-level interventions are more likely to promote greater gains (Wahler & Dumas, 1989; Wasik, Ramey, Bryant, & Sparling, 1990). Multilevel intervention may be particular critical when covariance is involved. Thus it is not surprising that long-term beneficial effects are more likely to occur for malnourished infants who participated in intervention programs involving both supplemental feeding and psychosocial stimulation (Grantham-McGregor, Gardner, Walker, & Powell, 1990).

2. Gains also are more likely to occur when interventions are tailored to the context within which the child is functioning. Available evidence suggests that interventions that do not fit the cultural patterns within which the child is reared are less likely to be effective. For example, traditional teacher-focused educational classrooms do not appear to be particularly effective for Hawaiian children, given the strong emphasis on peer relations within the Hawaiian culture; when classroom procedures were changed to incorporate more peer interaction, Hawaiian children's learning increased substantially (Tharp, 1989). Context implies more than just culture, however. The importance of mesosystem context is noted in the point made by Campbell (in press), that the use of lenient controls at home interferes with the ability of hyperactive children to develop internalized controls in school. Context also may reflect within-level influences, as seen in the report by Bronfenbrenner (1974) indicating that cognitive gains resulting from parent training programs tend to be extremely limited when the child lives in a crowded home environment. In this situation a more appropriate intervention strategy might be use of center-based intervention rather than home-based parent training. The critical

point is that it is not sufficient simply to be aware of contextual factors; rather interventions must be designed to build upon existing contextual factors.

3. In designing interventions it may be important to take advantage of the operation of *active covariance*. The principle of active covariance specifies that children are likely to seek out preferred "niches," with the characteristics of the preferred niche depending upon the child's individual characteristics and prior experience. Evidence with normal children (Carbo, 1983), and with children having severe developmental disorders (Kolka, Anderson, & Campbell, 1980), indicates that these children tend to perform better when they are in a preferred environment. This suggests the importance of providing a variety of *microenvironmental contexts*, which would allow children to select their preferred niches (Benninger, 1978; Wachs, 1987c). For example, working within an architectural design framework, Shaw (1987) has emphasized the importance of developing three-dimensional playground spaces, encompassing both open stages and "defensible spaces" (places where a child can observe group activities before deciding whether or not to participate). Looked at in terms of active covariance, we would predict that open stages would be preferred niches for more socially oriented children, who could use these stages as a place to perform and obtain attention from other peers ("look at me"). In contrast defensible spaces might be highly preferred niches for temperamentally shy children, who may need this type of space when faced with a crowded, active playground situation.

4. Not all children are created equal: To the extent that organism-environment interactions are operative, we should not expect all children to respond equally well to the same intervention. Some illustrations of this principle seem almost too obvious. It is not surprising to learn that cognitive behavior therapy appears to work best for children who have reached the formal operations stage (Durlak, Fuhrman, & Lampman, 1991), or that moderately hyperactive children are more likely to be able to use delayed positive reinforcement to help control their behavior than are extremely hyperactive children (Hersher, 1985). The potential relevance of organism-environment interaction to the design of

intervention projects is not, however, restricted only to the obvious. Evidence is available indicating that highly structured intervention programs work better with disadvantaged preschoolers than with handicapped preschoolers, whereas the reverse holds for program intensity and duration (Casto, 1987). Providing increased sensory stimulation may be particularly inappropriate for preterms, who may not yet have the sensory or cognitive capacity to adequately process increased stimulation (Turkewitz & Kenny, 1985); rather, a reduction in stimulation may be more appropriate for this population (Fajardo, Browning, Fisher, & Patton, 1990).[1]

Even when direct evidence is not available, the existence of certain individual characteristics points to the potential need for different types of intervention programs. As one possibility, Shotwell, Wolf, and Gardner (1980) have noted that some children seem mainly interested in objects and object manipulation, whereas other children seem primarily interested in social interaction and social events. To the extent that these differences between children are stable, it would be logical to assume (and test) whether socially oriented children will learn better when exposed to the types of sib or adult guided participation experiences described by Rogoff (1990), or whether object-oriented children will do better when left more or less alone with a variety of interesting objects. A similar argument also could be made in regard to differences in activity level. Based on evidence cited in Chapter 8, it may well be that low active children will benefit more when provided with interventions that involve adult mediation with the child, whereas high active children may benefit more from interventions that allow for greater self-discovery by the child. Rather than assuming that the same intervention program will work equally well for all children, it may be important to build in matches between children's characteristics and program characteristics to take advantage of the operation of organism-environment interaction.

5. Environmental intervention should be used to build in stress buffers. Research on resilient children has repeatedly noted that these children often appear to have personal characteristics that enable them to resist the detrimental impact of later stress. While some of these characteristics may reflect individual differences in

temperament (e.g., activity level, sociability; Werner & Smith, 1982), even biologically rooted temperament characteristics can be influenced by subsequent characteristics of the environment (Matheny, Wilson, & Thoben, 1987; Wachs, 1988c). Further, there also appear to be a number of individual characteristics that function as buffers, which are much more likely to be directly rooted in early patterns of caregiver child relations. These include self-esteem (Rutter, 1985c) and secure attachment (Sroufe & Egeland, 1991). To the extent that we can develop intervention programs oriented toward facilitating development of these individual buffers, we may be able to promote greater resilience when children are faced with later environmental stresses. Given the multidetermined nature of these characteristics, interventions designed to promote the development of buffers will have to focus not only on the child but also on the caregiver and the context within which the caregiver functions.

The Nature of Nurture: A Personal Statement

Over the past decade there have been major gains, both in our understanding of the structure of the environment and in our understanding of the processes underlying environment-development relations. For those researchers who love parsimony, it may seem that we have retreated rather than advanced, in the sense that our models of environmental action are getting to be more rather than less complex. While they are indeed more complex, our models also more accurately reflect both the nature of the environment and the processes of environmental action. What I would hope for during the next decade of study on environmental influences is not just a refinement of existing models of environmental action or more theory-driven research. Rather I believe that, for progress to be maintained, it is equally critical to change the way we approach research on environmental influences.

For the most part those of us involved in the study of environmental influences have operated in isolation from colleagues in different yet potentially important disciplines, such as behavior genetics, anthropology, or nutrition. Indeed, even within our own discipline, researchers studying the role of environmental influences with a specific age group all too often have been operating with little input from environmental researchers who use different age groups; the same isolation holds for environmental researchers studying human and infrahuman populations. Obviously, like the environment, this point is bidirectional; colleagues in different disciplines or areas have operated in relative isolation as well. Given the structure and nature of the environment, as detailed in this volume, it seems critical that the research process should mirror the nature of the environment. We cannot hope to fully understand the operation or the structure of the environment without input and collaboration from researchers in other relevant disciplines. Again the converse also holds; researchers in other disciplines cannot hope to fully understand the role of biological, temporal, or cultural influences upon development without input from environmental researchers. While there are many obstacles to cross-discipline collaboration (or even within-discipline collaboration), there also appears to be a greater awareness among researchers in different disciplines about the necessity for this type of collaboration (e.g., from the viewpoint of behavior genetics, see Plomin, 1990; from the viewpoint of nutrition, see Dewey & Schurch, 1990). I hope these isolated cries for greater cross-discipline collaboration will eventually become actualized in the form of joint research projects on the environmental system.

A similar point also can be made in regard to the application of basic research on environmental influences to real world problems. There still continues to be a tremendous gap between basic environmental researchers and intervention specialists. Yet perhaps nowhere else is such collaboration needed. Basic researchers can help design intervention programs that are congruent with what is known about environmental action, thus maximizing the chances for more successful intervention. Intervention specialists can provide environmental researchers with "real life" problems, as a means of testing the validity and boundaries of environmental action models. As one example I previously noted the potential importance of providing a *variety* of available niches for children. While architectural designers are particularly skilled in understanding the nature of space and the relation of space and objects, they are less well trained in observing and categorizing children's characteristics. Environmental researchers, while less skilled in their understanding of the interrelation of space and object, are more likely to be able to accurately classify children's characteristics. A collaboration between environmental researchers and architectural designers, on the question of what types of specific niches would be most appropriate for different types of children, would seem to be one obvious means of designing more effective interventions.

I hope that the concept of the environmental system will serve to inspire greater collaboration among researchers from different domains, and between researchers and intervention specialists, around the common problems of what determines variability in developmental outcomes and what can we do to promote optimal outcomes for individual children.

Note

1. It is important to keep in mind that use of terms like *highly structured* or *reduced stimulation* do not refer to absolute values. Rather these terms refer to the values associated with a specific program. A level of structure (or stimulation) that is labeled as high in one program may be labeled as low in another program (Bradley, personal communication). Thus, when I talk about increasing or decreasing intensity or structure, I am referring to change *relative* to the values in a specific intervention program.

References

Abraham, K., Kuehl, R., & Christopherson, V. (1983). Age specificity influences of parental behaviors upon the development of empathy in preschool children. *Child Study Journal, 13*, 175-185.

Ackerman, B. (1986). The relation between attention to the incidental context and memory for words in children and adults. *Journal of Experimental Child Psychology, 41*, 149-183.

Ackerman, P., Anhalt, J., Holcomb, T., & Dykman, R. (1986). Presumably innate and acquired automatic processes in children with attention and/or reading disorders. *Journal of Child Psychology and Psychiatry, 27*, 513-529.

Aiello, J., Nicosia, G., & Thompson, D. (1979). Physiological, social and behavioral consequences of crowding on children and adolescents. *Child Development, 10*, 195-202.

Ainsworth, M., & Bell, S. (1970). Attachment exploration and separation. *Child Development, 41*, 49-66.

Almli, C., & Finger, S. (1987). Neural insult and critical period concepts. In M. Bornstein (Ed.), *Sensitive periods in development*. Hillsdale, NJ: Lawrence Erlbaum.

Altmann, J. (1974). Observational study of behavior: Sampling methods. *Behavior, 39*, 227-267.

Anthony, E. (1978). Theories of change and children at-risk for change. In E. Anthony & C. Chiland (Eds.), *The child and his family* (Vol. 5). New York: John Wiley.

Aslin, R. (1981). Experiential influences and sensitive periods in perceptual development: Unified model. In R. Aslin, J. Albert, & M. Petersen (Eds.), *Development of perception*. New York: Academic Press.

Baird, J., Baglivi, G., & Kane, J. (1986). *The chronicles of Salimbene de Adam.* Binghampton, NY: Center for Medieval and Early Renaissance Studies.

Bakeman, R., Adamson, L., Konner, M., & Barr, R. (1990). Kung infancy: The social context of object exploration. *Child Development, 61,* 794-809.

Bakeman, R., & Brown, J. (1980). Early interaction: Consequences for social and mental development at 3 years. *Child Development, 51,* 437-447.

Barglow, P., Vaughn, B., & Molitor, N. (1987). Effects of maternal absence due to employment on the quality of infant mother attachment in a low risk sample. *Child Development, 58,* 945-954.

Barnard, K., & Bee, H. (1983). The impact of temporally patterned stimulation on the development of preterm infants. *Child Development, 54,* 1156-1167.

Baron, R., & Kenny, D. (1986). The moderator-mediator variable distinction in social psychological research. *Journal of Personality and Social Psychology, 11,* 1173-1182.

Barron, A., & Earls, F. (1984). The relation of temperament and social factors to behavior problems in three year old children. *Journal of Child Psychology and Psychiatry, 25,* 23-33.

Bates, J. (1989). Application of temperament concepts. In G. Kohnstamm, J. Bates, & M. Rothbart (Eds.), *Temperament in childhood.* New York: John Wiley.

Bateson, P. (1979). How do sensitive periods arise and what are they for. *Animal Behavior, 27,* 470-476.

Bateson, P. (1983). Rules and reciprocity in behavioral development. *Journal of Child Psychology and Psychiatry, 24,* 11-18.

Bateson, P., & Hinde, R. (1987). Developmental changes in sensitivity to experience. In M. Bornstein (Ed.), *Sensitive periods in development.* Hillsdale, NJ: Lawrence Erlbaum.

Beaton, G. (1986). Toward harmonization of dietary, biochemical and clinical assessments. *Nutrition Reviews, 44,* 349-358.

Beckwith, L., & Parmelee, A. (1986). EEG patterns of preterm infants, home environment and later IQ. *Child Development, 57,* 777-789.

Bee, H., Barnard, K., Eyres, S., Gray, C., Hamond, N., Spietz, A., Snyder, C., & Clark, B. (1982). Prediction of IQ and language skills from parental status, child performance, family characteristics and mother infant interaction. *Child Development, 53,* 1134-1156.

Beit-Hallahmi, B. (1987). Critical periods in psychoanalytic theories of personality development. In M. Bornstein (Ed.), *Sensitive periods in development.* Hillsdale, NJ: Lawrence Erlbaum.

Bell, R., & Chapman, M. (1986). Child effects and studies using experimental or brief longitudinal approaches to socialization. *Developmental Psychology, 22,* 595-603.

Belmont, J., & Freeseman, L. (1988). *Journal references to Jean Piaget and Lev Vygotsky: Implications for the culture of American psychology.* Unpublished manuscript, University of Kansas Medical Center, Kansas City.

Belsky, J. (1980). Mother infant interaction at home and in the laboratory: A comparative study. *Journal of Genetic Psychology, 137,* 37-47.

Belsky, J. (1984). The determinants of parenting: A process model. *Child Development, 55,* 83-96.

Belsky, J. (1988). The effect of infant daycare reconsidered. *Early Childhood Research Quarterly, 3*, 235-272.

Belsky, J. (1990). Parental and nonparental child care and children's socioemotional development. *Journal of Marriage and the Family, 52*, 885-903.

Belsky, J., & Rovine, M. (1988). Nonmaternal care in the first year of life and the security of infant parent attachment. *Child Development, 59*, 157-167.

Belsky, J., Rovine, M., & Taylor, D. (1984). The Pennsylvania Infant and Family Development Project III: The origins of individual differences in infant mother attachment. *Child Development, 55*, 718-728.

Benjamin, J. (1961). The innate and the experiential in child development. In H. Brosin (Ed.), *Lectures in experimental psychiatry*. Pittsburgh, PA: University of Pittsburgh Press.

Benjamin, J. (1965). Developmental biology and psychoanalysis. In N. Greenfield & W. Lewis (Eds.), *Psychoanalysis and current biological thought*. Madison: University of Wisconsin Press.

Benninger, C. (1978). Situation and setting: A search for an optimal environment. In E. Anthony & J. Chiland (Eds.), *The child and his family* (Vol. 5). New York: John Wiley.

Berg, K. (1975). Cardiac components of the defensive response in infants. *Psychophysiology, 12*, 224.

Berg, K., & Smith, M. (1983). Behavioral thresholds for tones during infancy. *Journal of Experimental Child Psychology, 35*, 409-425.

Bergman, L. (1981). Is intellectual development more vulnerable in boys than in girls? *Journal of Genetic Psychology, 138*, 175-181.

Bergman, P., & Escalona, S. (1949). Unusual sensitivity in very young children. *Psychoanalytic Study of the Child, 3*, 333-352.

Billings, A., & Moos, R. (1983). Comparisons of children of depressed and non-depressed parents. *Journal of Abnormal Child Psychology, 11*, 463-486.

Block, J., & Block, J. (1981). Studying situational dimensions. In D. Magnusson (Ed.), *Toward a psychology of situations*. Hillsdale, NJ: Lawrence Erlbaum.

Block, J., Block, J., & Gjerde, P. (1986). The personality of children prior to divorce. *Child Development, 57*, 827-740.

Block, J., Block, J., & Keyes, S. (1988). Longitudinally foretelling drug usage in adolescence: Early childhood personality and environmental precursors. *Child Development, 59*, 336-355.

Bloom, B. (1964). *Stability and change in human characteristics*. New York: John Wiley.

Bohlin, G., Hagekull, B., Germer, M., Andersson, K., & Lindberg, L. (1989). Avoidant and resistant reunion behaviors as predicted by maternal interactive behavior and infant temperament. *Infant Behavior and Development, 12*, 105-118.

Bolter, J., & Long, C. (1985). Methodological issues in research in developmental neuropsychology. In L. Hartlidge & C. Telzrow (Eds.), *The neuropsychology of individual differences*. New York: Plenum.

Boring, E. (1957). *A history of experimental psychology* (2nd ed.). New York: Appleton-Century-Crofts.

Bornstein, M. (1985). How infant and mother jointly contribute to developing cognitive competence in the child. *Proceedings of the National Academy of Science, 82*, 7470-7073.

Bornstein, M. (1989a). Cross-cultural developmental comparisons. *Developmental Review, 9,* 171-204.

Bornstein, M. (1989b). Sensitive periods in development. *Psychological Bulletin, 105,* 179-197.

Bornstein, M. (1989c). Between caretakers and their young. In M. Bornstein & J. Bruner, *Interaction in human development.* Hillsdale, NJ: Lawrence Erlbaum.

Bornstein, M., Azuma, H., Tamis-LeMonda, C., & Ogino, M. (1990). Mother and infant activity and interaction in Japan and in the United States: I. A comparative macroanalysis of naturalistic exchanges. *International Journal of Behavioral Development, 13,* 267-288.

Bornstein, M., & Tamis-LeMonda, C. (1990). Activities and interactions of mothers and their first born infants in the first six months of life. *Child Development, 61,* 1206-1217.

Bradley, R., & Caldwell, B. (1984). The relation of infants' home environment to achievement test performance in first grade. *Child Development, 55,* 803-809.

Bradley, R., Caldwell, B., & Rock, S. (1988). Home environment and school performance: A ten year follow-up and examination of three models of environmental action. *Child Development, 59,* 852-867.

Bradley, R., Caldwell, B., Rock, S., Casey, P., & Nelson, J. (1987). The early development of low birth weight infants. *International Journal of Behavioral Development, 10,* 301-318.

Bradley, R., Caldwell, B., Rock, S., Ramey, C., Barnard, K., Gray, C., Hammond, M., Mitchell, S., Gottfried, A., Siegel, L., & Johnson, D. (1989). Home environment and cognitive development in the first three years of life: A collaborative study involving six sites and three ethnic groups in North America. *Developmental Psychology, 25,* 217-235.

Bradley, R., Rock, S., Whiteside, L., Caldwell, B., Ramey, C., Barnard, K., Gray, C., Hammond, M., Mitchell, S., Gottfried, A., Siegel, L., & Johnson, D. (1990, April). *Early home environment and mental test performance: A structural analysis.* Paper presented at the meeting of the International Society on Infant Studies, Montreal.

Breitmayer, B., & Ramey, C. (1986). Biological nonoptimality and quality of postnatal environment as codeterminants of intellectual development. *Child Development, 57,* 1151-1165.

Brimer, E., & Levine, F. (1983). Stimulus seeking in hyperactive and non-hyperactive children. *Journal of Abnormal Child Psychology, 11,* 131-140.

Bronfenbrenner, U. (1974). *Is early education effective?* (Publication #0HD76-30025). Washington, DC: Department of Health, Education and Welfare.

Bronfenbrenner, U. (1977). Toward an experimental ecology of human development. *American Psychologist, 32,* 513-531.

Bronfenbrenner, U. (1986). Ecology of the family as a context for human development. *Developmental Psychology, 22,* 723-742.

Bronfenbrenner, U. (1989). Ecological system theories. *Annals of Child Development, 6,* 187-249.

Bronfenbrenner, U. (1991, July). *The developing ecology of human development.* Paper presented at the meeting of the International Society for the Study of Behavioral Development, Minneapolis.

Bronfenbrenner, U. (in press). The ecology of cognitive development. In R. Wozniak & K. Fisher (Eds.), *Specific environments: Thinking in context*. Hillsdale, NJ: Lawrence Erlbaum.

Bronfenbrenner, U., & Crouter, A. (1983). The evolution of environmental models in developmental research. In W. Kessen (Ed.), *Handbook of child psychology* (Vol. 1, 4th ed.). New York: John Wiley.

Bronson, W. (1985). Growth in the organization of behavior in the second year of life. *Developmental Psychology, 21*, 1108-1117.

Brooks-Gunn, J., & Warren, M. (1989). Biological and social contributions to negative affect in young adolescent girls. *Child Development, 60*, 40-55.

Burke, J., Moccia, P., Borus, J., & Burns, B. (1986). Emotional distress in 5th grade children 10 months after a natural disaster. *Journal of the American Academy of Child Psychiatry, 26*, 536-541.

Cahan, S., & Cohen, N. (1989). Age versus schooling effects on intelligence development. *Child Development, 60*, 1239-1249.

Campbell, S. (in press). The socialization and social development of hyperactive children. In M. Lewis & S. Miller (Eds.), *Handbook of developmental psychopathology*. New York: Plenum.

Canter, D. (1977). *The psychology of place*. London: Architectural Press.

Carbo, M. (1983). Research on reading and learning style. *Exceptional Children, 49*, 486-494.

Casto, G. (1987). Plasticity and the handicapped child. In J. Gallagher & C. Ramey (Eds.), *The malleability of children*. Belmont, CA: Brooks/Cole.

Caudill, W., & Weinstein, H. (1969). Maternal care and infant behavior in Japan and America. *Psychiatry, 32*, 12-43.

Chase-Lansdale, P., & Owen, M. (1987). Maternal employment in a family context. *Child Development, 58*, 1505-1512.

Chavez, A., & Martinez, C. (1984). Behavioral measurements of activity in children and their relation of food intake in a poor community. In E. Pollitt & P. Amante (Eds.), *Energy intake and activity*. New York: Liss.

Chein, I. (1954). The environment as a determinant of behavior. *Journal of Social Psychology, 39*, 115-127.

Chess, S., & Thomas, A. (1989). The practical application of temperament to psychiatry. In W. Carey & S. McDevitt (Eds.), *Clinical and educational applications of temperament research*. Amsterdam: Swets & Zeitlinger.

Chisholm, J. (1981). Prenatal influences on aboriginal-white Australian differences in neonatal irritability. *Ethology and Sociobiology, 2*, 67-73.

Chisholm, J. (1983). *Navaho infancy*. New York: Aldine.

Christie, D., & Glickman, C. (1980). The effects of classroom noise on children. *Psychology in the Schools, 17*, 405-408.

Clarke, A., & Clarke, A. (1979). Early experience: Its limited effect upon late development. In D. Shaffer & J. Dunn (Eds.), *The first year of life*. New York: John Wiley.

Clarke, A., & Clarke, A. (1986). Thirty years of child psychology. *Journal of Child Psychology and Psychiatry, 27*, 719-759.

Clarke, A., & Clarke, A. (1989). The later cognitive effects of early intervention. *Intelligence, 13*, 289-297.

Clarke-Stewart, K. (1978). And daddy makes three: The father's impact on mother and young child. *Child Development, 49*, 446-478.

Clarke-Stewart, K. (1989). Infant daycare: Maligned or malignant. *American Psychologist, 44*, 266-273.

Cloninger, C., Bohman, M., & Sigvardsson, S. (1981). The inheritance of alcohol abuse. *Archives of General Psychiatry, 38*, 861-868.

Cloninger, C., Sigvardsson, S., Bohman, M., & VonKnorring, A. (1982). Predisposition to petty criminality in Swedish adoptees. *Archives of General Psychiatry, 39*, 1242-1249.

Coates, D., & Lewis, M. (1984). Early mother infant interaction and infant cognitive status as predictors of school performance and cognitive behavior in six year olds. *Child Development, 55*, 1219-1230.

Cohen, J. (1988). *Statistical power analysis for the behavioral sciences* (2nd ed.). Hillsdale, NJ: Lawrence Erlbaum.

Cohen, J., & Cohen, P. (1983). *Applied multiple regression correlation analyses for the behavioral sciences* (2nd ed.). Hillsdale, NJ: Lawrence Erlbaum.

Cohen, S., Evans, G., Stokols, D., & Krantz, D. (1986). *Behavior, health and environmental stress*. New York: Plenum.

Cohen, S., Parmelee, A., Sigman, M., & Beckwith, L. (1982). Neonatal risk factors in preterm infants. *Applied Research in Mental Retardation, 3*, 262-278.

Colombo, J. (1982). The critical period concept. *Psychological Bulletin, 91*, 260-275.

Compas, B. (1987). Coping with stress during childhood and adolescence. *Psychological Bulletin, 101*, 393-403.

Compas, B., Slavin, L., Wagner, B., & VanNatta, K. (1986). Relationship of life events and social support with psychological dysfunction among adolescents. *Journal of Youth and Adolescence, 15*, 205-221.

Conger, J., Moisan-Thomas, P., & Conger, A. (1989). Cross situational generalizability of social competence. *Behavioral Assessment, 11*, 411-432.

Corno, L., Mitman, A., & Hedges, L. (1981). The influence of direct instruction on students' self appraisal: A hierarchical analysis of treatment and aptitude treatment interaction effects. *American Educational Research Journal, 18*, 39-61.

Cotterell, J. (1986). Work and community influences and the quality of child rearing. *Child Development, 57*, 362-347.

Cravioto, J., & DeLicarde, E. (1972). Environmental correlates of severe clinical malnutrition and language development in survivors from kwashiorkor or marasmus. In *Nutrition, the nervous system and behavior* (PAHO Scientific Publication 251). Washington, DC: PAHO.

Crnic, K., & Greenberg, M. (1990). Minor parenting stresses with young children. *Child Development, 61*, 1682-1637.

Crnic, K., Greenberg, M., Ragozin, R., Robinson, N., & Bashman, R. (1983). The effects of stress and social support on mothers and premature and full term infants. *Child Development, 54*, 209-217.

Crockenberg, S. (1981). Infant irritability, mother responsiveness and social support influences on the security of infant mother attachment. *Child Development, 52*, 857-865.

Crockenberg, S. (1986). Are temperamental differences in babies associated with predictable differences in caregivers? In J. Lerner & R. Lerner (Eds.), *Temperament and psychosocial interaction in children*. San Francisco: Jossey-Bass.

Crockenberg, S. (1987). Predictors and correlates of anger toward and punitive control of toddlers by adolescent mothers. *Child Development, 58*, 964-975.

Crockenberg, S., & Littman, C. (1990). Autonomy is competence in two year olds: Maternal correlates of child compliance, noncompliance and self-assertion. *Developmental Psychology, 26*, 961-971.

Crockenberg, S., & McCluskey, K. (1986). Change in maternal behavior during the baby's first year of life. *Child Development, 57*, 746-753.

Cronbach, L. (1991). Emerging views on methodology. In T. D. Wachs & R. Plomin (Eds.), *Conceptualization and measurement of organism-environment interaction.* Washington, DC: American Psychological Association.

Cronbach, L., & Snow, R. (1977). *Aptitudes and instructional methods.* New York: Halstead.

Crouter, A., MacDermid, S., McHale, S., & Perry-Jenkins, M. (1990). Parental monitoring and perceptions of children's school performance and conduct in dual and single earner families. *Developmental Psychology, 26*, 649-657.

Crouter, A., & McHale, S. (in press). The long arm of the job: Influences of parental child rearing. In T. Luster & L. Okagaki (Eds.), *Parenting: An ecological perspective.* Hillsdale, NJ: Lawrence Erlbaum.

Crouter, A., Perry-Jenkins, M., Huston, T., & McHale, S. (1987). Processes underlying father involvement in dual-earner and single-earner families. *Developmental Psychology, 23*, 431-440.

Crowell, J., & Feldman, S. (1988). Mothers' internal models of relationships and children's behavioral and developmental status. *Child Development, 59*, 1273-1285.

Cummings, E. (1987). Coping with background anger in early childhood. *Child Development, 58*, 976-984.

Cummings, E., Iannotti, R., & Zahn-Waxler, C. (1985). Influence of conflict between adults on the emotions and aggression of young children. *Developmental Psychology, 21*, 495-507.

Cummings, E., Vogel, G., Cummings, J., & El-Sheikh, M. (1989). Children's responses to different forms of expression of anger between adults. *Child Development, 60*, 1392-1404.

Cummings, E., Zahn-Waxler, C., & Radke-Yarrow, M. (1981). Young children's responses to expression of anger and affection by others in the family. *Child Development, 52*, 1274-1282.

Cummings, J., Pellegrini, D., Notarius, C., & Cummings, E. (1989). Children's responses to angry adult behavior as a function of marital distress and history of interparent hostility. *Child Development, 60*, 1035-1043.

Cutrona, C., & Troutman, B. (1986). Social support, infant temperament and parenting self-efficacy. *Child Development, 57*, 1507-1518.

Daniels, D., Dunn, J., Furstenberg, F., & Plomin, R. (1985). Environmental differences within the family and adjustment differences within pairs of adolescent siblings. *Child Development, 56*, 764-774.

David, T., & Weinstein, C. (1987). The built environment and children's development. In C. Weinstein & T. David (Eds.), *Spaces for children.* New York: Plenum.

Davidson, G. (1980). Child rearing as a response to traditional aboriginal and modern Western value systems. In E. Anthony & C. Chiland (Eds.), *The child and his family* (Vol. 6). New York: John Wiley.

Day, R., & Ghandour, M. (1984). The effects of television mediated aggression and real life aggression on the behavior of Lebanese children. *Journal of Experimental Child Psychology, 38,* 7-18.

Deal, J., Halverson, C., & Wampler, K. (1989). Parental agreement on child rearing orientation: Relations to parental, marital, family and child characteristics. *Child Development, 60,* 1025-1034.

DeVries, M. (1984). Temperament and infant mortality among the Masai of East Africa. *American Journal of Psychiatry, 141,* 1189-1194.

Dewey, K., & Schurch, B. (1990). Needs and priorities for research and action from the behavioral point of view. In B. Schurch & N. Scrimshaw (Eds.), *Activity, energy expenditure and energy requirements of infants and children.* Lausanne, Switzerland: IDECG.

Dixon, S., LeVine, R., Richman, R., & Brazelton, T. (1984). Mother child interaction around a teaching task. *Child Development, 55,* 1255-1264.

Dodge, K. (1986). Social information processing variables in the development of aggression and altruism in children. In C. Zahn-Waxler, E. Cummings, & R. Iannotti (Eds.), *Altruism and aggression.* Cambridge: Cambridge University Press.

Dodge, K., Coie, J., Pettit, G., & Price, J. (1990). Peer status and aggression in boys groups. *Child Development, 61,* 1289-1309.

Dodge, K., Pettit, G., McCluskey, C., & Brown, M. (1980). Social competence in children. *Monographs for the Society for Research in Child Development, 51*(213).

Doherty, W., & Needle, R. (1991). Psychological adjustment and substance use among adolescents before and after a parental divorce. *Child Development, 62,* 328-337.

Dornbush, S., Ritter, P., Leiderman, P., Roberts, D., & Fraleigh, M. (1987). The relation of parenting style to adolescent school performance. *Child Development, 58,* 1244-1257.

Douglas, J. (1975). Early hospital admission and later disturbance of behavior and learning. *Developmental Medicine and Child Neurology, 17,* 456-480.

Draper, P. (1973). Crowding among hunter gathers. *Science, 182,* 301-303.

Drillien, C. (1964). *The growth and development of the prematurely born infant.* Edinburgh, UK: Livingstone.

Dumaret, A. (1985). IQ, scholastic performance and behavior of sibs raised in contrasting environments. *Journal of Child Psychology and Psychiatry, 26,* 553-580.

Dunkin, N., & Doenau, S. (1980). A replication study of unique and joint contribution to variance in student achievement. *Journal of Educational Psychology, 72,* 394-403.

Dunn, J., Kendrick, C., & MacNamee, R. (1981). The reaction of first born children to the birth of a sibling. *Journal of Child Psychology and Psychiatry, 22,* 1-18.

Dunn, J., & Plomin, R. (1990). *Separate lives: Why siblings are so different.* New York: Basic Books.

Durlak, J., Fuhrman, T., & Lampman, C. (1991). Effectiveness of cognitive-behavior therapy for maladapting children. *Psychological Bulletin, 110*, 204-214.

Dykman, R., Ackerman, P., Holcomb, B., & Boudreau, Y. (1983). Physiological manifestation of learning disabilities. *Journal of Learning Disabilities, 16*, 46-53.

Edelbrook, C., Costello, A., Dulcan, M., Kalas, R., & Conover, N. (1985). Age differences in the reliability of the psychiatric interview of the child. *Child Development, 56*, 265-275.

Egeland, B., & Farber, E. (1984). Infant mother attachment: Factors related to its development and change over time. *Child Development, 55*, 753-777.

Elder, G., Downey, G., & Cross, C. (1986). Family ties and life changes. In N. Datan, A. Green, & H. Reese (Eds.), *Lifespan developmental psychology*. Hillsdale, NJ: Lawrence Erlbaum.

Elder, G., Van Nguyen, T., & Caspi, A. (1985). Linking family hardship to children's lives. *Child Development, 56*, 361-375.

Emery, R. (1982). Interparental conflict and the children of divorce and discord. *Psychological Bulletin, 92*, 310-330.

Emery, R., Weintraub, S., & Neale, J. (1982). Effect of marital discord on the school behavior of children of schizophrenic, affectively disordered and normal parents. *Journal of Abnormal Child Psychology, 10*, 215-228.

Engle, P. (1990). *Child caregiving and infant preschool nutrition.* Paper prepared for the Cornell Food and Nutrition Policy Project and the Rockefeller Foundation, Polytechnic State University, San Luis Obispo, CA.

Enns, J., & Akhter, M. (1989). A developmental study of filtering and visual attention. *Child Development, 60*, 1188-1199.

Epstein, H. (1976). A biologically based framework for intervention projects. *Mental Retardation, 14*, 26-27.

Erlenmeyer-Kimling, L., & Cornblatt, B. (1987). High risk research in schizophrenia. *Journal of Psychiatric Research, 21*, 401-411.

Escalona, S. (1968). *The roots of individuality.* Chicago: Aldine.

Evans, G., Kliewer, W., & Martin, J. (in press). The role of the physical environment and the health and well being of children. In H. Schroder (Ed.), *New directions in health psychology*. New York: Hemisphere.

Fabiani, M., Sohner, H., Tait, C., & Bordieri, O. (1984). Mathematical expression of relationship between auditory brain stem transmission time and age. *Developmental Medicine and Child Neurology, 26*, 461-465.

Fajardo, B., Browning, M., Fisher, D., & Patton, J. (1990). Effect of nursery environment on state regulation in very low birth weight premature infants. *Infant Behavior and Development, 13*, 287-303.

Farber, E., & Egeland, B. (1982). Developmental consequences of out of home care for infants in a low income population. In E. Zigler & E. Gordon (Eds.), *Daycare: Scientific and social policy issues*. Boston: Auburn House.

Fenderich, M., Warner, V., & Weissman, M. (1990). Family risk factors, parental depression and psychopathology in offspring. *Developmental Psychology, 26*, 40-50.

Field, T. (1981). Gaze behavior of normal and high risk infants during early interactions. *Journal of the American Academy of Child Psychiatry, 20*, 308-317.

Field, T., & Ignatoff, T. (1981). Videotaping effects on the behaviors of low income mothers and their infants during floor play interaction. *Journal of Applied Developmental Psychology, 2,* 227-236.

Fields, R. (1979). Child terror victims and adult terrorism. *Journal of Psychohistory, 7,* 71-75.

Fiese, B. (1990). Playful relationships: A contextual analysis of mother toddler interaction in symbolic play. *Child Development, 61,* 1648-1656.

Finkelstein, M., Gallagher, J., & Farran, D. (1980). Attentiveness and responsiveness to auditory stimuli of children at risk for mental retardation. *American Journal of Mental Deficiency, 85,* 135-144.

Forehand, R., McCombs, A., Long, M., Brody, G., & Fauber, R. (1988). Early adolescent adjustment to recent parental divorce. *Journal of Consulting and Clinical Psychology, 56,* 624-627.

Fowler, W. (1983). *Potentials of childhood* (Vol. 1). Lexington, MA: Lexington.

Frankel, F., Simmons, J., Fichter, M., & Freeman, B. (1984). Stimulus overselectivity and mentally retarded children. *Journal of Child Psychology and Psychiatry, 25,* 147-155.

Frankel, K., & Bates, J. (1990). Mother toddler problem solving: Antecedents in attachment, home behavior and temperament. *Child Development, 61,* 801-819.

Fraser, B., & Fisher, D. (1983). Student achievement as a function of person environment fit. *British Journal of Educational Psychology, 55,* 89-99.

Freebody, P., & Tirre, W. (1985). Achievement outcomes of two reading programs. *British Journal of Educational Psychology, 55,* 53-60.

Freedman, D. (1974). *Human infancy: An evolutionary perspective.* Hillsdale, NJ: Lawrence Erlbaum.

Fry, D. (1988). Intercommunity differences in aggression among Zapotec children. *Child Development, 59,* 1108-1019.

Fry, D., & Scher, A. (1984). The effects of father absence on children's achievement motivation, ego strength and locus of control orientation. *British Journal of Developmental Psychology, 2,* 167-178.

Fuller, J. (1967). Experiential deprivation and later behavior. *Science, 158,* 1645-1652.

Gamble, T., & Zigler, E. (1986). The effects of infant daycare. *American Journal of Orthopsychiatry, 56,* 26-42.

Gandour, M. (1989). Activity level as a dimension of temperament in toddlers: Its relevance for the organismic specificity hypothesis. *Child Development, 60,* 1092-1098.

Garmezy, N. (1983). Stressors of childhood. In N. Garmezy & M. Rutter (Eds.), *Stress, coping and development in children.* New York: McGraw-Hill.

Garmezy, N. (1987). Stress, competence and development. *American Journal of Orthopsychiatry, 57,* 159-174.

Gersten, R. (1983). Stimulus overselectivity in autistic, trainable mentally retarded and nonhandicapped children. *Journal of Abnormal Child Psychology, 11,* 61-76.

Gilmore, J., Best, H., & Eakins, S. (1980). Coping with test anxiety: Individual differences in seeking complex play materials. *Canadian Journal of Behavioral Science, 12,* 241-254.

Golden-Meadow, S., & Nylander, C. (1984). Gestural communication in deaf children. *Monographs of the Society for Research in Child Development, 49*(207), 3, 4.

Goodnow, J. (1988). Parents' ideas, actions and feelings. *Child Development, 59,* 286-302.

Goodnow, J., Cashmore, J., Cotton, S., & Knight, R. (1984). Mothers' developmental time tables in two cultural groups. *International Journal of Psychology, 19,* 193-205.

Gordon, B. (1981). Child temperament and adult behavior. *Child Psychiatry and Human Development, 11,* 167-178.

Gordon, B. (1983). Maternal perception of child temperament and observed mother child interactions. *Child Psychiatry and Human Development, 13,* 153-167.

Gottfried, A. (1984). *Home environment and early cognitive development.* New York: Academic Press.

Gottfried, A., & Gottfried, A. (1984). Home environment and cognitive development in young children of middle socioeconomic status families. In A. Gottfried (Ed.), *Home environment and early cognitive development.* New York: Academic Press.

Gottfried, A., & Gottfried, A. (1989, April). *Home environment and children's academic intrinsic motivation.* Paper presented at the meeting of the Society for Research in Child Development, Kansas City.

Gottlieb, G. (1991). Experiential canalization of behavioral development: Theory. *Developmental Psychology, 27,* 4-13.

Graham, P. (1977). Environmental influences on psychosocial development. *International Journal of Mental Health, 6,* 7-31.

Graham, P., & Canavan, K. (1982). The mental health of children in developing countries. In E. Anthony & C. Chiland (Eds.), *The child and his family* (Vol. 7). New York: John Wiley.

Grantham-McGregor, S., Gardner, J., Walker, S., & Powell, C. (1990). The relationship between undernutrition, activity level, and development in young children. In B. Schurch & N. Scrimshaw (Eds.), *Activity, energy expenditure and energy requirements of infant and children.* Lausanne, Switzerland: IDECG.

Grantham-McGregor, S., Schofield, W., & Powell, C. (1987). Development of severely malnourished children who receive psychosocial stimulation. *Pediatrics, 79,* 247-254.

Greenacre, P. (1952). *Trauma, growth and personality.* New York: Norton.

Greenberg, M., & Crnic, K. (1988). Longitudinal predictors of developmental status and social interaction in premature and full-term infants at age 2. *Child Development, 59,* 554-570.

Greenberger, E., Steinberg, L., & Vaux, A. (1982). Person environment congruence as predictors of adolescent health and behavioral problems. *American Journal of Community Psychology, 10,* 511-525.

Greenough, W., Black, J., & Wallace, C. (1987). Experience and brain development. *Child Development, 58,* 539-559.

Greenough, W., & Juraska, J. (1979). Experience induced changes in brain fine structure: Their behavioral implications. In M. Hahn, C. Gensen, & B. Dudek (Eds.), *Development and evolution in brain size: Behavioral implications.* New York: Academic Press.

Grych, J., & Fincham, F. (1990). Marital conflict and children's adjustment. *Psychological Bulletin, 108,* 267-290.

Gunnar, M. (1978). Changing a frightening toy into a pleasant toy by allowing the infant to control its action. *Developmental Psychology, 14,* 157-162.

Gunnar, M., & Donahue, M. (1980). Sex difference in social responsiveness between 6 months and 12 months. *Child Development, 51*, 262-265.

Gurdinbeldi, J., & Perry, J. (1985). Divorce and mental health sequelae for children. *Journal of the American Academy of Child Psychiatry, 24*, 531-537.

Hambrick-Dixon, P. (1986). The effects of experimentally imposed noise on task performance of black children attending daycare centers near elevated subway trains. *Developmental Psychology, 22*, 259-264.

Hambrick-Dixon, P. (1988). The effect of elevated subway train noise over time on black children's visual vigilance performance. *Journal of Environmental Psychology, 8*, 299-314.

Hamilton, J. (1983). Development of interest and enjoyment in adolescents. *Journal of Youth and Adolescence, 12*, 355-362.

Hammen, C., Adrian, C., Gordon, D., Burge, D., Jaenicke, C., & Hiroto, D. (1987). Children of depressed mothers: Maternal strain and symptom predictors of dysfunction. *Journal of Abnormal Psychology, 96*, 190-198.

Hannan, K., & Luster, T. (1990, April). *Influence of child, parent and contextual factors on quality of home environment*. Paper presented at the 7th International Conference on Infant Studies, Montreal.

Hanson, R. (1975). Consistency and stability of home environmental measures related to IQ. *Child Development, 46*, 470-480.

Harkness, S., & Super, C. (1982). Why African children are so hard to test. In L. Adler (Ed.), *Cross-cultural research at issue*. New York: Academic Press.

Hartup, W. (1983). Peer relations. In E. Hetherington (Ed.), *Handbook of child psychology* (Vol. 4). New York: John Wiley.

Haskins, R. (1985). Public school aggression among children with varying daycare experience. *Child Development, 56*, 689-703.

Heckhausen, J. (1987a). How do mothers know? Infants' chronological age or infants' performance as determinants of adaptation and maternal instruction. *Journal of Experimental Child Psychology, 43*, 212-226.

Heckhausen, J. (1987b). Balancing for weakness and challenging developmental potential. *Developmental Psychology, 23*, 762-770.

Henderson, N. (1980). The effects of early experience on the behavior of animals. In E. Simmel (Ed.), *Early experience and later behavior*. New York: Academic Press.

Hersher, L. (1985). The effectiveness of behavior modification on hyperkinesis. *Child Psychiatry and Human Development, 16*, 87-96.

Hess, R., & McDevitt, T. (1984). Some cognitive consequences of maternal intervention techniques. *Child Development, 55*, 2017-2023.

Hetherington, E. (1989). Coping with family transition: Winners, losers and survivors. *Child Development, 60*, 1-14.

Hetherington, M., Cox, M., & Cox, R. (1979). Play and social interaction in children following divorce. *Journal of Social Issues, 35*, 26-49.

Hetherington, M., Cox, M., & Cox, R. (1985). Long term effects of divorce and remarriage on the adjustment of children. *Journal of the American Academy of Child Psychiatry, 24*, 518-530.

Hinde, R., & Stevenson-Hinde, J. (1973). *Constraints on learning*. New York: Academic Press.

Ho, D., & Kang, T. (1984). Intergenerational comparison of child rearing attitudes and practices in Hong Kong. *Developmental Psychology, 20,* 1004-1006.

Hock, E., & Clinger, J. (1980). Behavior toward mother and stranger of infants who have experienced group daycare, individual care or exclusive maternal care. *Journal of Genetic Psychology, 137,* 49-61.

Hoffman, L. (1989). Effects of maternal employment in the two parent family. *American Psychologist, 44,* 283-292.

Hoffman, L. (1991). The influence of the family environment on personality. *Psychological Bulletin, 110,* 187-203.

Holden, G., & Edwards, L. (1989). Parental attitudes toward child rearing: Instruments, issues and implications. *Psychological Bulletin, 106,* 29-58.

Holden, G., & West, M. (1989). Proximate regulation by mothers. *Child Development, 60,* 64-69.

Horowitz, F. (1987). *Exploring developmental theories.* Hillsdale, NJ: Lawrence Erlbaum.

Howes, C. (1990). Can the age of entry into child care and the quality of child care predict adjustment in kindergarten? *Developmental Psychology, 26,* 292-303.

Howes, C., & Olenick, M. (1986). Family and child care influences upon toddlers compliance. *Child Development, 57,* 202-216.

Howes, C., & Stewart, P. (1987). Child's play with adults, toys and peers: An examination of family and child care influences. *Developmental Psychology, 23,* 423-430.

Hubert, N., & Wachs, T. D. (1985). Parental perceptions of the behavioral components of infant easiness-difficultness. *Child Development, 56,* 1525-1537.

Hunt, J. M. (1961). *Intelligence and experience.* New York: Ronald.

Hunt, J. M. (1977). *Specificity in early development and experience.* O'Neill Invited Lecture, Meyer Children's Rehabilitation Institute, University of Nebraska Medical Center.

Hunt, J. M. (1979). Psychological development: Early experience. *Annual Review of Psychology, 30,* 103-143.

Hyson, M. (1983). Going to the doctor: A developmental study of stress and coping. *Journal of Child Psychology and Psychiatry, 24,* 247-259.

Ide, J., Parkerson, J., Haertal, G., & Walberg, H. (1981). Peer group influences on educational outcomes. *Journal of Educational Psychology, 73,* 473-494.

Irwin, R., Stillman, J., & Schade, A. (1986). The width of the auditory filter in children. *Journal of Experimental Child Psychology, 41,* 429-442.

Isabella, R., & Belsky, J. (1991). International synchrony and the origins of infant mother attachment. *Child Development, 62,* 373-384.

Itard, J. (1962). *The wild boy of Averon.* New York: Appleton-Century-Crofts.

Iverson, B., & Walberg, H. (1982). Home environment and school learning: A quantitative synthesis. *Journal of Experimental Education, 50,* 144-151.

Jacob, T. (1975). Family interaction in disturbed and normal families. *Psychological Bulletin, 82,* 33-65.

Jennings, K., & Connors, R. (1989). Mothers' interactional style and children's competence at 3 years. *International Journal of Behavioral Development, 12,* 155-175.

Joffe, J. (1982). Approaches to prevention of adverse developmental consequences of genetic and prenatal factors. In L. Bond & J. Joffe (Eds.), *Facilitating infant and early childhood development.* Hanover: University Press of New England.

Kagan, J. (1967). On the need for relativism. *American Psychologist, 22,* 131-142.

Kagan, J. (1984). *The nature of the child.* New York: Basic Books.

Kagitcibasi, C., & Berry, J. (1989). Cross-cultural psychology: Current research and trends. *Annual Review of Psychology, 40,* 493-532.

Kail, R., & Bisanz, J. (1982). Cognitive development: An information processing perspective. In R. Vasta (Ed.), *Strategies and techniques of child study.* New York: Academic Press.

Kaplan, H., & Dove, H. (1987). Infant development among the Ache of Eastern Paraguay. *Developmental Psychology, 23,* 190-198.

Kawasaki, T., Delea, C., Barter, F., & Smith, A. (1978). The effect of high sodium and low sodium intakes on blood pressure and other related variables in human subjects with idiopathic hypertension. *American Journal of Medicine, 64,* 193-198.

Kendler, K., & Eaves, L. (1986). Models from joint effects of genotype on environment on liability to psychiatric illness. *American Journal of Psychiatry, 143,* 279-299.

Keogh, B. (1986a). The future of the learning disability field. *Journal of Learning Disabilities, 19,* 455-460.

Keogh, B. (1986b). Temperament and schooling: Meaning of goodness of fit. In J. Lerner & R. Lerner (Eds.), *Temperament and social interaction in infants and children.* San Francisco: Jossey-Bass.

Kindermann, T., & Skinner, E. (1988). Developmental tasks as organizers of children's ecologies. In J. Valsiner (Ed.), *Children's development within socially culturally structured environments.* Norwood, NJ: Ablex.

Klingman, A., & Wiesner, E. (1982). The relationship of proximity to tension areas and size of settlement to fear level of Israel children. *Journal of Behavior Therapy and Experimental Psychology, 13,* 321-323.

Koch, H. (1956a). Attentiveness of young children towards peers is related to certain characteristics of their siblings. *Psychological Monographs, 70*(426).

Koch, H. (1956b). Sissyness and tomboyishness in relation to sibling characteristics. *Journal of Genetic Psychology, 88,* 231-244.

Koch, H. (1956c). Some emotional attitudes of the young child in relation to characteristics of a sibling. *Child Development, 27,* 393-426.

Kochanska, G. (1990). Maternal beliefs as long term predictors of mother-child interaction and report. *Child Development, 61,* 1934-1943.

Kochanska, G., Kuczynski, L., & Radke-Yarrow, M. (1989). Correspondence between mother's self reported and observed child rearing practices. *Child Development, 60,* 56-73.

Kogan, N. (1983). Stylistic variation in childhood and adolescence. In J. Flavell & E. Markman (Eds.), *Mussen's handbook of child psychology* (1st ed.). New York: John Wiley.

Kolka, D., Anderson, L., & Campbell, M. (1980). Sensory preference and over selective responding in autistic children. *Journal of Autism and Developmental Disorders, 10,* 259-271.

Korn, S., & Gannon, S. (1983). Temperament, cultural variation and behavior disorder in preschool children. *Child Psychiatry and Human Development, 13,* 203-212.

Kuczynski, L., Kochanska, G., Radke-Yarrow, M., & Girnius-Brown, O. (1987). A developmental interpretation of young children's noncompliance. *Developmental Psychology, 23,* 799-806.

Lamb, B., & Hwang, C. (1982). Maternal attachment and maternal neonate bonding. In M. Lamb & A. Brown (Eds.), *Advances in developmental psychology* (Vol. 2). Hillsdale, NJ: Lawrence Erlbaum.

Lane, D., & Pearson, D. (1982). The development of selective attention. *Merrill Palmer Quarterly, 28,* 317-337.

Langmeier, J., & Matejcek, Z. (1975). *Psychological deprivation in childhood.* New York: John Wiley.

Lee, C., & Bates, J. (1985). Mother child interaction at age 2 years and perceived difficult temperament. *Child Development, 56,* 1314-1345.

Lempers, J., Clark-Lempers, D., & Simons, R. (1989). Economic hardship, parenting and distress in adolescence. *Child Development, 60,* 25-49.

Leng, Y., & Ong, S. (1982). Malaysian child psychology practices. *Australian and New Zealand Journal of Psychiatry, 16,* 61-66.

Lerner, R., Lerner, J., Windle, M., Hooker, K., Lenerz, K., & East, P. (1986). Children and adolescents in their contexts: Test of a goodness of fit model. In R. Plomin & J. Dunn (Eds.), *The study of temperament: Changes, continuity and challenges.* Hillsdale, NJ: Lawrence Erlbaum.

Lester, B., & Brazelton, T. (1982). Cross-cultural assessment of neonatal behavior. In D. Wagner & H. Stevenson (Eds.), *Cultural perspectives on child development.* San Francisco: Freeman.

LeVine, R., Miller, P., & Richman, A. (1991, April). *Influence of culture and schooling on mother's models of infant care.* Paper presented at the meeting of the Society for Research in Child Development, Seattle, WA.

Levine, S. (1962). Psychophysiological effects of infant stimulation. In E. Bliss (Ed.), *Roots of behavior.* New York: Harper & Row.

Levy-Shiff, R. (1982). The effects of father absence on young children in mother headed families. *Child Development, 53,* 1400-1405.

Lewin, K. (1936). *Principles of topological psychology.* New York: McGraw-Hill.

Lewis, M., & Feiring, C. (1989). Infant mother and mother infant interaction behavior and subsequent attachment. *Child Development, 60,* 831-837.

Lewis, M., Feiring, C., McGuffog, C., & Jaskir, J. (1984). Predicting psychopathology in six year olds from early social relations. *Child Development, 55,* 123-136.

Liben, L. (1981). Spatial representation and behavior. In L. Liben, A. Patterson, & N. Newcombe (Eds.), *Spatial representation and behavior across the lifespan.* New York: Academic Press.

Lieh-Mak, F. (1980). Boat children: The implication of high density living. In E. Anthony & C. Chiland (Eds.), *The child and his family.* New York: John Wiley.

Lin, C., & Fu, V. (1990). A comparison of child rearing practices among Chinese, immigrant Chinese and Caucasian American parents. *Child Development, 61,* 429-433.

Loo, C. (1972). The effects of spatial density on the social behavior of children. *Journal of Applied Social Psychology, 2* 372-381.

Lounsbury, M., & Bates, J. (1982). The cries of infants of differing levels of perceived temperamental difficultness. *Child Development, 52,* 677-686.

Lumley, M., Ables, L., Melamed, B., Pistone, L., & Johnson, J. (1990). Coping outcome in children undergoing stressful medical procedures: The role of child-environment variables. *Behavioral Assessment, 12,* 223-238.

Luster, T., & Dubow, E. (1991, April). *Home environment and maternal intelligence as predictors of verbal intelligence.* Paper presented at the meeting of the Society for Research in Child Development, Seattle, WA.

Luster, T., Rhodes, K., & Hass, B. (1989). The relation between parenting values and parenting behavior. *Journal of Marriage and the Family, 51,* 139-147.

Lytton, H. (1980). *Parent child interaction.* New York: Plenum.

Maccoby, E., Snow, M., & Jacklin, C. (1984). Children's disposition and mother-child interactions at 12 and 18 months. *Developmental Psychology, 20,* 459-472.

MacDonald, K. (1985). Early experience, relative plasticity and social development. *Developmental Review, 5,* 99-121.

MacDonald, K. (1986). Early experience, relative plasticity and cognitive development. *Journal of Applied Developmental Psychology, 7,* 101-124.

MacDonald, K., & Parke, R. (1984). Bridging the gap: Parent child play interaction and peer interactive competence. *Child Development, 55,* 1265-1277.

MacPhee, D., & Ramey, C. (1981, March). *Infant temperament as a catalyst and consequent of development in two caregiving environments.* Paper presented at the 14th Gatlinberg Conference on Research in Mental Retardation and Developmental Disabilities, Gatlinberg, TN.

Madden, J., O'Hara, J., & Levenstein, T. (1984). Home again: Effects of the mother child home program on mother and child. *Child Development, 56,* 636-647.

Magnusson, D. (1988). *Individual development from an interactional perspective.* Hillsdale, NJ: Lawrence Erlbaum.

Malhotra, S. (1989). Varying risk factors and outcomes: An Indian perspective. In W. Carey & S. McDevitt (Eds.), *Clinical and educational applications of temperament research.* Amsterdam: Swets & Zeitlinger.

Mann, J., Have, J., Plunkett, J., & Meisels, S. (1991). Time sampling: A methodological critique. *Child Development, 62,* 227-241.

Marjoribanks, K. (1981). Family environment and children's academic achievement. *Journal of Psychology, 109,* 155-164.

Martin, J. (1981). A longitudinal study of the consequences of early mother infant interaction. *Monograph of the Society for Research in Child Development, 190*(46, No. 3).

Masten, A. (1989). Resilience in development: Implications of the study of successful adaptation for developmental psychopathology. In D. Cicchetti (Ed.), *Rochester Symposium on Developmental Psychopathology* (Vol. 1). Hillsdale, NJ: Lawrence Erlbaum.

Matheny, A., Wilson, R., & Thoben, A. (1987). Home and mother: Relations with infant temperament. *Developmental Psychology, 23,* 323-331.

Maziade, M., Cote, R., Boutin, P., Bernier, H., & Thivierge, J. (1987). Temperament and intellectual development. *American Journal of Psychiatry, 144,* 144-150.

Maxwell, S. (1990). Why are interactions so difficult to detect? *Behavioral and Brain Sciences, 13,* 140-141.

McCall, R. (1977). Challenges to a science of developmental psychology. *Child Development, 48,* 333-344.

McCall, R. (1981). Nature-nurture and the two realms of development. *Child Development, 52,* 1-12.

McCall, R. (1983). Environmental effects on intelligence. *Child Development, 54,* 408-415.

McCall, R. (1984). Developmental changes in mental performance. *Child Development, 85,* 1317-1321.

McCall, R. (1991). So many interactions, so little evidence. Why? In T. D. Wachs & R. Plomin (Eds.), *Conceptualization and measurement of organism environment interaction.* Washington, DC: American Psychological Association.

McCall, R., Myers, E., Hartman, J., & Roche, A. (1983). Developmental changes in head circumference and mental performance growth rates. *Developmental Psychobiology, 16,* 457-468.

McCartney, K., Harris, M., & Bernieri, F. (1990). Growing up and growing apart: A developmental meta-analysis of twin studies. *Psychological Bulletin, 107,* 226-237.

McCord, R., & Wakefield, J. (1981). Arithmetic achievement as a function of introversion extroversion and teacher presented reward and punishment. *Personality and Individual Differences, 2,* 145-152.

McCullough, A., Kirksey, A., Wachs, T. D., McCabe, G., Bassily, N., Bishry, Z., Galal, O., Harrison, G., & Jerome, N. (1990). Vitamin B6 status of Egyptian mothers: Relation to infant behavior and maternal infant interactions. *American Journal of Clinical Nutrition, 51,* 1067-1074.

McGivern, J., & Levin, J. (1983). The key word method in children's vocabulary learning. *Contemporary Educational Psychology, 8,* 46-54.

McKeon, R. (1941). *The basic works of Aristotle.* New York: Random House.

McLoyd, V. (1983). The effects of the structure of play objects on the pretend play of low income preschool children. *Child Development, 54,* 626-638.

McLoyd, V. (1990). The impact of economic hardship on black families and children. *Child Development, 61,* 311-346.

McSwain, R. (1981). Care and conflict in infant development. *Infant Behavior and Development, 4,* 225-246.

McWhirter, L., & Trew, K. (1982). Children in Northern Ireland: A lost generation? In E. Anthony & C. Chiland (Eds.), *The child and its family* (Vol. 7). New York: John Wiley.

Mednick, S., Brennan, P., & Kandel, E. (1988). Predisposition to violence. *Aggressive Behavior, 14,* 25-33.

Mednick, S., Gabrielli, W., & Hutchings, B. (1984). Genetic influences in criminal condition: The evidence from an adoption cohort. *Science, 224,* 890-894.

Meyer-Probst, B., Rosler, H., & Teichmann, H. (1983). Biological and psychosocial risk factors and development during childhood. In D. Magnusson & V. Allen (Eds.), *Human development: An interactional perspective.* New York: Academic Press.

Milgram, S. (1970). The experience of living in cities. *Science, 167,* 1461-1468.

Milich, R., & Dodge, K. (1984). Social information processing in child psychiatric populations. *Journal of Abnormal Child Psychology, 12,* 471-490.

Millar, S. (1984). Reasons for the attribution of intent in 7 and 9 year old children. *British Journal of Developmental Psychology, 2,* 51-61.

Miller, L., & Bizzell, R. (1983). Long-term effects of four preschool programs. *Child Development, 54,* 727-741.

Miller, S. (1988). Parents' beliefs about children's cognitive development. *Child Development, 59,* 259-285.

Mink, I., & Nihira, K. (1986). Family lifestyles and child behaviors: A study of direct effects. *Developmental Psychology, 22,* 610-616.

Mischel, W. (1973). Toward a cognitive social learning reconceptualization of personality. *Psychological Review, 80,* 252-283.

Moore, D., & Sheik, A. (1971). Toward a theory of early infantile autism. *Psychological Review, 78,* 451-456.

Moore, G. (1987). The physical environment and cognitive development in child care centers. In C. Weinstein & T. David (Eds.), *Spaces for children.* New York: Plenum.

Moorehouse, M. (1991). Linking maternal employment patterns to mother child activities and children's school competence. *Developmental Psychology, 27,* 295-303.

Moos, R. (1973). Conceptualization of human environment. *American Psychologist, 28,* 652-665.

Morice, R. (1980). Change in the aboriginal adolescent. In E. Anthony & C. Chiland (Eds.), *The child and its family.* New York: John Wiley.

Moskovitz, S. (1985). Longitudinal follow-up of child survivors of the Holocaust. *Journal of the American Academy of Child Psychiatry, 24,* 401-407.

Munsinger, H. (1975). The adopted child's IQ: A critical review. *Psychological Bulletin, 82,* 623-659.

Murphy, L., & Moriarty, A. (1976). *Vulnerability, coping and growth.* New Haven, CT: Yale University Press.

Myers, B. (1987). Mother-infant bonding as a critical period. In M. Bornstein (Ed.), *Sensitive periods in development.* Hillsdale, NJ: Lawrence Erlbaum.

Neims, A. (1986). Individuality in the response to dietary constituents: Some lessons from drugs. *Nutrition Review Supplements, 21,* 237-241.

Nihira, K., Mink, I., & Shapiro, C. (1991, April). *Home environment of developmentally disabled children.* Paper presented at the meeting of the Society for Research in Child Development, Seattle, WA.

Nihira, K., Tomiyasu, Y., & Oshio, C. (1987). Homes of TMR children: Comparison between American and Japanese families. *American Journal of Mental Deficiency, 91,* 486-495.

Novak, M., Olley, J., & Kearney, D. (1980). Social skills of children with special needs and integrated in separate preschools. In T. Field, S. Goldberg, D. Stern, & A. Sostek (Eds.), *High risk infants and children.* New York: Academic Press.

Nunberg, H. (1955). *Principles of psychoanalysis.* New York: International University Press.

O'Brien, M. (1988, April). *Cognitive and social predictors of early language.* Paper presented at the meeting of the International Conference on Infant Studies, Washington, DC.

O'Brien, M., Johnson, J., & Anderson-Goetz, D. (1989). Evaluating quality in mother infant interaction: Situational effects. *Infant Behavior and Development, 12,* 451-464.

Oehler, J., Eckerman, C., & Wilson, W. (1988). Social stimulation and the regulation of premature infants' states prior to term age. *Infant Behavior and Development, 11*, 333-351.

Olson, S., Bates, J., & Bayles, K. (1984). Mother infant interaction and the development of individual differences in children's cognitive competence. *Developmental Psychology, 20*, 166-179.

Olweus, D. (1986). Aggression and hormones. In D. Olweus, J. Block, & M. Radke-Yarrow (Eds.), *Development of antisocial and prosocial behavior*. New York: Academic Press.

Openshaw, D., Thomas, D., & Rollins, B. (1984). Parental influence on adolescent self-esteem. *Journal of Early Adolescence, 4*, 259-274.

Oyama, S. (1979). The concept of the sensitive period in developmental studies. *Merrill Palmer Quarterly, 25*, 83-103.

Parpal, M., & Maccoby, E. (1985). Maternal responsiveness and subsequent child compliance. *Child Development, 56*, 1326-1344.

Parrinello, R., & Ruff, H. (1988). The influence of adult intervention and infant's level of attention. *Child Development, 59*, 1125-1135.

Patterson, G. (1983). Stress: A change agent for family process. In N. Garmezy & M. Rutter (Eds.), *Stress, coping and development in children*. New York: McGraw-Hill.

Patterson, G., & Yoerger, K. (1991, April). *A model for general parenting skills is too simple: Mediational models work better*. Paper presented at the meeting of the Society for Research in Child Development, Seattle, WA.

Pavenstedt, E. (1967). *The drifters: Children of disorganized lower class families*. Boston: Little, Brown.

Pedersen, F., Rubenstein, J., & Yarrow, L. (1979). Infant development in father absent families. *Journal of Genetic Psychology, 135*, 51-61.

Perris, H., Myers, N., & Clifton, R. (1990). Long-term memory for a single infancy experience. *Child Development, 61*, 1796-1807.

Pettit, G., Dodge, J., & Brown, M. (1988). Early family experience, and social problem solving patterns and children's social competence. *Child Development, 59*, 107-120.

Phillips, D., McCartney, K., & Scarr, S. (1987). Child care quality and children's social development. *Developmental Psychology, 23*, 537-543.

Pianta, R., Sroufe, L., & Egeland, B. (1989). Continuity and discontinuity in maternal sensitivity at 6, 24 and 42 months in a high risk sample. *Child Development, 60*, 481-487.

Pickles, A., Rutter, M., & Quinton, D. (1989, April). *Statistical and conceptual models of turning points in developmental processes*. Paper presented at the Second ESF Workshop on Methodological Issues and Longitudinal Research, Oslow, Norway.

Plato. (1926). *The laws* (Vol. 1; R. Bury, Ed.). Cambridge, MA: Harvard University Press.

Plato. (1937). *The republic* (Vol. 1; P. Shorey, Ed.). Cambridge, MA: Harvard University Press.

Pliszka, S. (1989). Effect of anxiety on cognition, behavior and stimulant response in ADHD. *Journal of the American Academy of Child and Adolescent Psychiatry, 28*, 882-887.

Plomin, R. (1990). The role of inheritance in behavior. *Science, 248*, 183-188.

Plomin, R., & Bergeman, C. (1991). The nature of nurture: Genetic influence on environmental measures. *Behavioral and Brain Sciences, 14,* 373-428.

Plomin, R., & Daniels, D. (1984). The interaction between temperament and environment. *Merrill Palmer Quarterly, 30,* 149-162.

Plomin, R., & Daniels, D. (1987). Why are children in the same family so different from each other? *Behavior and Brain Sciences, 10,* 1-16.

Plomin, R., DeFries, J., & Loehlin, J. (1977). Genotype environment interaction and correlation in the analysis of human development. *Psychological Bulletin, 84,* 309-322.

Plomin, R., DeFries, J., & McClearn, G. (1980). *Behavior genetics: A primer.* San Francisco: Freeman.

Plomin, R., Loehlin, J., & DeFries, (1985). Genetic and environmental components of "environmental" influences. *Developmental Psychology, 21,* 391-402.

Plunkett, J., Berlin, M., Dedrick, C., Dichtelmiller, M., & Meisels, S. (1990, April). *Socioemotional adaptation of extremely low birth weight (ELBW) infants at 14 months: Home and laboratory observations.* Paper presented at the 7th International Conference on Infant Studies, Montreal.

Pollitt, E. (1983). Morbidity and infant development: An hypothesis. *International Journal of Behavior Development, 6,* 461-475.

Powell, C., & Grantham-McGregor, S. (1985). The ecology of nutritional status and development in young children in Kingston, Jamaica. *American Journal of Clinical Nutrition, 41,* 1322-1331.

Radke-Yarrow, M., & Sherman, T. (in press). Hard growing: Children who survive. In J. Rolf, A. Masten, D. Cicchetti, K. Neuchterlein, & S. Weintrubg (Eds.), *Risk and protective factors in the development of psychopathology.* Cambridge: Cambridge University Press.

Rahmanifar, A., Kirksey, A., Wachs, T. D., McCabe, G., Sobhy, A., Galal, O., Harrison, G., & Jerome, N. (1992). *Determinants of infant behavior and caregiver infant interaction in a semirural Egyptian population.* Manuscript submitted for publication.

Ramey, C., MacPhee, D., & Yeates, K. (1982). Preventing developmental retardation: A general systems model. In L. Bond & J. Joffe (Eds.), *Facilitating infant and early childhood development.* Hanover: University Press of New England.

Rauh, V., Achenbach, T., Nurcombe, B., Howell, C., & Teti, D. (1988). Minimizing adverse effects of low birth weight: Four year results of an early intervention. *Child Development, 59,* 544-553.

Reed, M., Pien, D., & Rothbart, M. (1984). Inhibitory self-control in preschool children. *Merrill Palmer Quarterly, 30,* 131-147.

Ricciuti, H., & Thomas, M. (1990, April). *Early maternal and environmental correlates of quality of infant care and 18 month Bayley performance.* Paper presented at the International Conference on Infant Studies, Montreal.

Roberts, W. (1986). Nonlinear models of development. *Child Development, 57,* 1167-1178.

Rogoff, B. (1990). *Apprenticeship in thinking: Cognitive development and social context.* New York: Oxford University Press.

Rogoff, B., Mosier, C., Mistry, J., & Goncu, A. (in press). Toddler guided participation in cultural activity. *Cultural Dynamics.*

Rose, R., Kaprio, J., Williams, C., Viken, R., & Obrenski, K. (1990). Social contact and sibling similarity. *Behavior Genetics, 20,* 763-778.

Rosenthal, R., & Lani, F. (1981). Selective attention and self-control in delinquent adolescents. *Journal of Youth and Adolescence, 10,* 211-220.

Rovee-Collier, C. (1984). The ontogeny of learning and memory in human infancy. In R. Kail & N. Spear (Eds.), *Comparative perspectives on the development of memory.* Hillsdale, NJ: Lawrence Erlbaum.

Rowe, G., & Plomin, R. (1981). The importance of nonshared environmental influences in behavioral development. *Developmental Psychology, 17,* 517-531.

Rushton, P., Brainerd, C., & Pressley, M. (1983). Behavioral development and construct validity: The principle of aggregation. *Psychological Bulletin, 94,* 18-38.

Rutter, M. (1979). Protective responses in children's responses to stress and disadvantage. In M. Kent & J. Rolf (Eds.), *Primary prevention of psychopathology* (Vol. 3). Hanover: University Press of New England.

Rutter, M. (1981a). Stress coping and development: Some issues and some questions. *Journal of Child Psychology and Psychiatry, 22,* 323-356.

Rutter, M. (1981b). Epidemiological-longitudinal strategies and causal research in child psychiatry. *Journal of the American Academy of Child Psychiatry, 20,* 513-544.

Rutter, M. (1983a). School effect on pupil progress: Research findings and policy implications. *Child Development, 54,* 1-29.

Rutter, M. (1983b). Statistical and personal interactions. In D. Magnusson & V. Allen (Eds.), *Human development: An interactional perspective.* New York: Academic Press.

Rutter, M. (1985a). Family and school influences on behavioral development. *Journal of Child Psychology and Psychiatry, 26,* 349-368.

Rutter, M. (1985b). Family and school influences on cognitive development. In R. Hinde, A. Parret-Cleremont, & J. Stevenson-Hinde (Eds.), *Social relationships in cognitive development.* Oxford: Clarendon.

Rutter, M. (1985c). Resilience in the face of adversity. *British Journal of Psychiatry, 147,* 598-611.

Rutter, M., & Garmezy, N. (1983). Developmental psychopathology. In E. Hetherington (Ed.), *Socialization, personality and social development: Vol. 4. Mussen's handbook of child psychology* (4th ed.). New York: John Wiley.

Rutter, M., & Pickles, A. (1991). Person environment interaction: Concepts, mechanism and implications for data analysis. In T. D. Wachs & R. Plomin (Eds.), *Conceptualization and measurement of organism environment interaction.* Washington DC: American Psychological Association.

Sackett, G. (1978). *Observing behavior* (Vol. 2). Baltimore: University Park Press.

Sackett, G. (1991). Toward a more temporal view of organism environment interaction and development. In T. D. Wachs & R. Plomin (Eds.), *Conceptualization and measurement of organism environment interaction.* Washington, DC: American Psychological Association.

Sackett, G., Ruppenthal, G., Fahrenbruch, C., Holm, R., & Greenough, W. (1981). Social isolation rearing effects in monkeys vary with genotype. *Developmental Psychology, 17,* 313-318.

Saco-Pollitt, C., Pollitt, E., & Greenfield, D. (1985). The cumulative deficit hypothesis in the light of cross-cultural evidence. *International Journal of Behavioral Development, 8*, 75-97.

Sameroff, A. (1983). Developmental systems: Context and evolution. In W. Kessen (Ed.), *Mussen's handbook of child psychology* (Vol. 1, 4th ed.). New York: John Wiley.

Sameroff, A., & Chandler, M. (1975). Reproductive risk and the continuum of care taking causality. In F. Horowitz (Ed.), *Review of child development research* (Vol. 4). Chicago: University of Chicago Press.

Sameroff, A., Seifer, R., Barocas, R., Zax, M., & Greenspan, S. (1987). Intelligence quotient scores of 4 year old children: Social environmental risk factors. *Pediatrics, 79*, 343-350.

Sanda, A. (1982). The Nigerian family in transition. In E. Anthony & C. Chiland (Eds.), *The child and its family* (Vol. 7). New York: John Wiley.

Scarr, S., & McCartney, K. (1983). How people make their own environments. *Child Development, 54*, 424-435.

Scarr, S., & McCartney, K. (1988). Far from home: An experimental evaluation of the mother child home program in Bermuda. *Child Development, 59*, 531-543.

Scarr, S., & Weinberg, R. (1983). The Minnesota Adoption Studies. *Child Development, 54*, 260-267.

Schaefer, E., & Edgerton, M. (1985). Parent and child correlates of parental modernity. In I. Sigel (Ed.), *Parental belief systems*. Hillsdale, NJ: Lawrence Erlbaum.

Schaffer, H. (1966). Activity level as a constitutional determinant of infantile reaction to deprivation. *Child Development, 37*, 595-602.

Schaffer, H., & Emerson, P. (1964). Patterns of responsivity to physical contact in early human development. *Journal of Child Psychology and Psychiatry, 5*, 1-13.

Schmidt, K., Solant, M., & Bridger, W. (1985). Electrodermal activity of undersocialized aggressive children. *Journal of Child Psychology and Psychiatry, 26*, 653-660.

Sciara, F. (1975). Effects of father absence on the educational achievement of urban black children. *Child Study Journal, 5*, 45-55.

Scott, J. (1979). Critical periods in organizational processes. In S. Falkner & J. Tanner (Eds.), *Human growth, neurobiology and nutrition*. New York: Plenum.

Scrimshaw, N., & Young, V. (1989). Adaptation to low protein and energy intakes. *Human Organization, 48*, 20-30.

Seginer, R. (1983). Parents' educational expectations in children's academic achievement. *Merrill Palmer Quarterly, 29*, 1-23.

Seitz, V., Rosenbaum, L., & Apfel, N. (1985). Effects of family support intervention: A ten year follow-up. *Child Development, 56*, 376-391.

Sells, S. (1963). Dimensions of stimulus situation which account for behavioral variance. In S. Sells (Ed.), *Stimulus determinants of behavior*. New York: Ronald.

Shaheen, S. (1984). Neuro-maturational and behavioral development: The case of childhood lead poisoning. *Developmental Psychology, 20*, 542-550.

Shand, N., & Kosawa, Y. (1985). Culture transmission: Caudell's model and alternative hypotheses. *American Anthropologist, 87*, 862-871.

Shapiro, A. (1974). Effects of family density and mother's education on preschoolers' motor skills. *Perceptual Motor Skills, 38*, 79-86.

Shaw, L. (1987). Designing playgrounds for able and disabled children. In C. Weinstein & T. David (Eds.), *Spaces for children*. New York: Plenum.

Shell, R., Roosa, M., & Eisenberg, N. (1991, April). *Family and child parent relation influences on children's coping strategies*. Paper presented at meeting of the Society for Research in Child Development, Seattle, WA.

Shepher, J. (1971). Mate selection among second generation kibbutz adolescents. *Archives of Sexual Behavior, 1*, 293-307.

Shotwell, J., Wolf, D., & Gardner, H. (1980). Styles of achievement in early symbol use. In M. Foster & S. Brandes (Eds.), *Symbol as sense*. New York: Academic Press.

Siegel, L. (1982). Reproductive, perinatal and environmental factors as predictors of the cognitive and language development of preterm and full term infants. *Child Development, 53*, 963-973.

Sigel, I. (1985). *Parental belief systems: The psychological consequences for children*. Hillsdale, NJ: Lawrence Erlbaum.

Simmeonsson, R. (1978). Social competence. In J. Wortis (Ed.), *Mental retardation and developmental disabilities* (Vol. 10). New York: Brunner-Mazel.

Simmons, R., Burgeson, R., Ford, S., & Blyth, D. (1987). The impact of cumulative change in early adolescence. *Child Development, 58*, 1220-1234.

Skeels, H., & Dye, H. (1939). A study of the effects of differential stimulation of mentally retarded children. *Proceedings of the American Association on Mental Deficiency, 44*, 114-136.

Slabach, E., Morrow, J., & Wachs, T. D. (1991). Questionnaire measurement of infant and child temperament: Current status and future directions. In J. Streleu & A. Angleitner (Eds.), *Explorations in temperament*. New York: Plenum.

Slade, A. (1987). Quality of attachment and early symbolic play. *Developmental Psychology, 23*, 78-85.

Slife, B., & Rychlak, J. (1983). Role of affect assessment in modeling aggressive behavior. *Journal of Personality and Social Psychology, 43*, 861-868.

Smith, L., & Hagen, V. (1984). Relationships between home environment and sensory motor development of Down syndrome and nonretarded infants. *American Journal of Mental Deficiency, 89*, 124-132.

Snow, C. (1987). Relevance of the notion of a critical period to language acquisition. In M. Bornstein (Ed.), *Sensitive periods in development*. Hillsdale, NJ: Lawrence Erlbaum.

Sophian, C. (1986). Early development and children's spatial monitoring. *Cognition, 22*, 61-88.

Spangler, G. (1989). Toddlers' everyday experiences as related to preceding mental and emotional disposition and their relationship to subsequent mental and motivational development. *International Journal of Behavioral Development, 12*, 285-303.

Spitz, R. (1965). *The first year of life*. New York: International Universities Press.

Sroufe, A. (1979). The coherence of individual development. *American Psychologist, 34*, 831-841.

Sroufe, A. (1983). Infant caregiving attachment and patterns of adaptation and competence. In M. Perlmutter (Ed.), *Minnesota Symposium on Child Psychology* (Vol. 16). Hillsdale, NJ: Lawrence Erlbaum.

Sroufe, A., & Egeland, B. (1991). Person and environment: Illustrations of interactions from a longitudinal study of development. In T. D. Wachs & R. Plomin (Eds.), *Conceptualization and measurement of organism environment interaction.* Washington, DC: American Psychological Association.

Sroufe, A., Egeland, B., & Kreutzer, T. (1990). The fate of early experience following developmental change. *Child Development, 61,* 1361-1373.

Sroufe, A., & Fleeson, J. (1986). Attachment and the construction of relationships. In W. Hartup & Z. Rubin (Eds.), *Relationships and development.* Hillsdale, NJ: Lawrence Erlbaum.

Steinberg, L., Mount, T., Lanborn, S., & Dornbusch, S. (1991). Authoritative parenting and adolescent adjustment across varied ecological niches. *Journal of Research on Adolescence, 1,* 19-36.

Stevenson, H., & Lee, S. (1990). Contexts of achievement. *Monographs of the Society for Research in Child Development, 55*(221).

Stevenson, H., Parker, T., Wilkinson, A., Bonnevaux, V., & Gonzalez, M. (1978). Schooling, environment and cognitive development: A cross-cultural study. *Monographs of the Society for Research in Child Development, 43*(175).

Strelau, J. (1983). *Temperament, personality and activity.* New York: Academic Press.

Super, C. (1976). Environmental effects on motor development. *Developmental Medicine and Child Neurology, 18,* 561-567.

Super, C. (1981, April). *Cultural construction of behavior problems in infancy.* Symposium presentation for the Society for Research in Child Development, Boston.

Super, C., & Harkness, S. (1986). Temperament, development and culture. In R. Plomin & J. Dunn (Eds.), *The study of temperament: Changes, continuities and challenges.* Hillsdale, NJ: Lawrence Erlbaum.

Super, C., Herrera, M., & Mora, J. (1990). Long term effects of food supplementation and psychosocial intervention on the physical growth of Colombian infants at risk of malnutrition. *Child Development, 61,* 29-49.

Sutcliffe, J. (1980). On the relationship of reliability to statistical power. *Psychological Bulletin, 88,* 509-515.

Tachibana, Y., & Hasegawa, E. (1986). Aggressive responses of adolescents to a hypothetical frustrating situation. *Psychological Reports, 58,* 111-118.

Tamis-LeMonda, C., & Bornstein, M. (in press). Language, play and attention at one year. *Infant Behavior and Development.*

Teti, D., Bond, L., & Gibbs, E. (1988). Mothers, fathers and siblings: A comparison of play styles and their influence upon infant cognitive level. *International Journal of Behavior Development, 11,* 415-432.

Thanaphum, S. (1980). Understanding and ensuring normal adolescent development in Thailand. In E. Anthony & C. Chiland (Eds.), *The child and its family.* New York: John Wiley.

Tharp, R. (1989). Psychocultural variables and constants: The effects on teaching and learning in school. *American Psychologist, 44,* 349-359.

Thomas, A., & Chess, S. (1991). Temperament and the concept of goodness of fit. In J. Strelau & A. Angleitner (Eds.), *Explorations in temperament.* New York: Plenum.

Thomas, N., & Berk, L. (1981). Effects of school environments on the development of young children's creativity. *Child Development, 52,* 1153-1162.

Thompson, W., & Grusec, J. (1970). Studies of early experience. In P. Mussen (Ed.), *Carmichael's manual of child psychology*. New York: Plenum.

Tienari, P., Sorri, A., Lahti, I., Naarala, M., Wahlberg, K., Ronkiko, J., Pohjla, J., & Moring, J. (1985). The Finnish adoptive family study of schizophrenia. *Yale Journal of Biology and Medicine, 58*, 227-327.

Torun, B. (1991). Short and long-term effects of low or restricted energy intakes on the activity of infants and children. In B. Schurch & N. Scrimshaw (Eds.), *Activity, energy expenditure and energy requirements of infants and children*. Lausanne, Switzerland: IDECG.

Trickett, P. (1983). The interaction of cognitive style and classroom environment in determining first grade behavior. *Journal of Applied Developmental Psychology, 4*, 43-65.

Turkewitz, G., & Kenny, P. (1985). The role of developmental limitation of sensory input on sensory perceptual organization. *Developmental and Behavioral Pediatrics, 6*, 302-306.

Turkheimer, M., Bakeman, R., & Adamson, L. (1989). Do mothers support and peers inhibit skilled object play in infants? *Infant Behavior and Development, 12*, 37-44.

Van Alstyne, D. (1929). *The environment of three year old children*. Unpublished doctoral dissertation, Columbia University, New York.

Vaughn, B., Block, J., & Block, J. (1988). Parental agreement on child rearing during early childhood and the psychological characteristics of adolescents. *Child Development, 59*, 1020-1033.

Vaux, A., & Ruggiero, M. (1983). Stressful life change and delinquent behavior. *American Journal of Community Psychology, 11*, 169-183.

Vibbert, M., & Bornstein, M. (1989). Specific associations between domains of mother child interactions and toddler referential language and pretense play. *Infant Behavior and Development, 12*, 163-184.

Volling, B., & Belsky, J. (1991). Multiple determinants of father involvement during infancy in dual-earner and single-earner families. *Journal of Marriage and the Family, 13*, 461-474.

Vygotsky, L. (1978). *Mind in society*. Cambridge, MA: Harvard University Press.

Waber, D., Vuori-Christiansen, L., Ortez, N., Clement, J., Christiansen, N., Mora, J., Reed, T., & Herrera, M. (1981). Nutritional supplementation, maternal education and cognitive development of infants at risk of malnutrition. *American Journal of Clinical Nutrition, 34*, 807-813.

Wachs, T. D. (1979). Proximal experience and early cognitive intellectual development: The physical environment. *Merrill Palmer Quarterly, 25*, 3-41.

Wachs, T. D. (1983). The use and abuse of environment in behavior genetic research. *Child Development, 54*, 396-407.

Wachs, T. D. (1984). Proximal experience and early cognitive-intellectual development: The social environment. In A. Gottfried (Ed.), *Home environment and early cognitive development*. New York: Academic Press.

Wachs, T. D. (1986). Models of physical environmental action. In A. Gottfried (Ed.), *Play interactions: The contribution of play material and parent involvement to child development*. New York: Lexington.

Wachs, T. D. (1987a). Short term stability of aggregated and nonaggregated measures of parent behavior. *Child Development, 58*, 796-797.

Wachs, T. D. (1987b). Specificity of environmental action as manifest in environmental correlates of infants' mastery motivation. *Developmental Psychology, 23*, 782-790.

Wachs, T. D. (1987c). Developmental perspectives on designing for development. In C. Weinstein & T. David (Eds.), *Spaces for children*. New York: Plenum.

Wachs, T. D. (1988a). Environmental assessment with developmentally disabled infants and preschoolers. In T. D. Wachs & R. Sheehan (Eds.), *Assessment of young developmentally disabled children*. New York: Plenum.

Wachs, T. D. (1988b). The validity of observer ratings of ambient background noise in the home. In U. Berglund, J. Berglund, J. Karlson, & T. Lindvall (Eds.), *Noise as a public health problem* (Vol. 3). Stockholm: Swedish Counsel.

Wachs, T. D. (1988c). Relevance of physical environment influences for toddler temperament. *Infant Behavior and Development, 11*, 431-445.

Wachs, T. D. (1989). The nature of the physical microenvironment: An expanded classification system. *Merrill Palmer Quarterly, 35*, 399-402.

Wachs, T. D. (1990). The development of effective child care environments. *Children's Environment Quarterly, 6*, 4-7.

Wachs, T. D. (1991a). Environmental considerations in the study of organism-environment interaction with nonextreme groups. In T. D. Wachs & R. Plomin (Eds.), *Conceptualization and measurement of organism environment interaction*. Washington, DC: American Psychological Association.

Wachs, T. D. (1991b). A shaky alliance: Commentary on Plomin and Bergeman's model of nature and nurture. *Behavioral and Brain Sciences, 14*, 411-412.

Wachs, T. D. (1991c). Temperament, activity and behavioral development of infants and children. In B. Schurch & N. Scrimshaw (Eds.), *Activity, energy expenditure and energy requirements of infants and children*. Lausanne, Switzerland: IDECG.

Wachs, T. D. (1991d). Conceptualization and measurement of organism-environment interactions: Synthesis and conclusions. In T. D. Wachs & R. Plomin (Eds.), *Conceptualization and measurement of organism environment interaction*. Washington, DC: American Psychological Association.

Wachs, T. D. (in press). Determinants of intellectual development: Single determinant research in a multidetermined universe. *Intelligence*.

Wachs, T. D., Bishry, A., Sobhy, A., & McCabe, G. (1991, April). *Sex differences in the relation of rearing environment to adaptive behavior of Egyptian toddlers*. Paper presented at the meeting of the Society for Research in Child Development, Seattle, WA.

Wachs, T. D., Bishry, Z., Sobhy, A., McCabe, G., Shaheen, F., & Galal, O. (1992). *Relation of rearing environment to cognitive performance and adaptive behavior of Egyptian toddlers*. Manuscript submitted for publication.

Wachs, T. D., & Camli, O. (1991). Do ecological or individual characteristics mediate the influence of the physical environment upon maternal behavior? *Journal of Environmental Psychology, 11*, 249-264.

Wachs, T. D., & Chan, A. (1986). Specificity of environmental action as seen in physical and social environment correlates of three aspects of twelve month infants' communication performance. *Child Development, 57*, 1464-1475.

Wachs, T. D., & Desai, S. (1992, May). *Environment, temperament and attachment at 24 months of age*. Paper presented at the meeting of the International Society for Infant Studies, Miami.

Wachs, T. D., & Gandour, M. J. (1983). Temperament, environment and six month cognitive-intellectual development. *International Journal of Behavioral Development, 6*, 135-152.

Wachs, T. D., & Gruen, G. (1982). *Early experience and human development.* New York: Plenum.

Wachs, T. D., Moussa, W., Bishry, Z., Yunis, F., Sobhy, A., McCabe, G., Jerome, N., Galal, O., Harrison, G., & Kirksey, A. (in press). Relations between nutrition and cognitive performance in Egyptian toddlers. *Intelligence.*

Wachs, T. D., & Plomin, R. (1991). *Conceptualization and measurement of organism environment interaction.* Washington, DC: American Psychological Association.

Wachs, T. D., Sigman, M., Bishry, Z., Moussa, W., Jerome, N., Neumann, C., Bwibo, N., & McDonald, M. (in press). Caregiver child interaction patterns in two cultures in relation to nutrition. *International Journal of Behavioral Development.*

Wade, B. (1981). Highly anxious pupils in formal and informal primary classrooms. *British Journal of Educational Psychology, 51*, 39-41.

Wagner, D. (1978). Memories of Morocco: The influence of age, schooling and environment on memory. *Cognitive Psychology, 10*, 1-28.

Wahler, R., & Dumas, J. (1989). Attentional problems in dysfunctional mother child interactions. *Psychological Bulletin, 105*, 116-130.

Wahlsten, D. (1990). Insensitivity of the analysis of variance to heredity-environment interaction. *Behavioral and Brain Sciences, 13*, 109-161.

Walden, T., & Baxter, A. (1989). The effect of context and age on social referencing. *Child Development, 60*, 1511-1518.

Walker, E., Downey, G., & Bergman, A. (1989). The effects of parental psychopathology and maltreatment on child behavior. *Child Development, 60*, 15-24.

Walker, E., & Emory, E. (1983). Infants at risk for psychopathology: Offspring of schizophrenic parents. *Child Development, 54*, 1269-1285.

Wasik, B., Ramey, C., Bryant, D., & Sparling, J. (1990). A longitudinal study of two earlier intervention studies. *Child Development, 61*, 1682-1696.

Wasserman, T. (1984). The effect of cognitive development on the use of cognitive behavioral techniques with children. *Child and Family Behavior Therapy, 5*, 37-50.

Webb, N. (1989). Peer interaction and learning in small groups. *International Journal of Educational Research, 13*, 21-37.

Weinstein, C. (1987). Designing preschool classrooms to support development. In C. Weinstein & T. David (Eds.), *Spaces for children.* New York: Plenum.

Weinstein, C., & David, T. (1987). *Spaces for children.* New York: Plenum.

Weisner, J. (1981). Cities, stress and children. In R. Munroe, R. Munroe, & B. Whiting (Eds.), *Handbook of cross cultural human development.* New York: Garland.

Weisner, T., Bernstein, M., Garnier, H., Rosenthal, J., & Hamilton, C. (1990, April). Children in conventional and nonconventional family lifestyle classified as in attachment at 12 months. Paper presented at the International Conference on Infant Study, Montreal.

Weisner, T., & Wilson-Mitchell, J. (1990). Nonconventional family lifestyles and sex typing in six year olds. *Child Development, 61*, 1915-1933.

Weisz, J., Rothbaum, F., & Blackburn, T. (1984). Standing out and standing in: The psychology of control in America and Japan. *American Psychologist, 39*, 955-969.

Weizmann, F. (1971). Correlational statistics and the nature nurture problem. *Science, 171*, 589.

Werner, E. (1979). *Cross-cultural child development*. Monterey, CA: Brooks/Cole.

Werner, E., & Smith, R. (1982). *Vulnerable but invincible*. New York: McGraw-Hill.

Werner, J., & Lipsitt, L. (1981). The infancy of human sensory systems. In E. Gollin (Ed.), *Developmental plasticity*. New York: Academic Press.

Whitehall, M., DeMyer-Gapin, S., & Scott, T. (1976). Stimulus seeking in antisocial preadolescent children. *Journal of Abnormal Psychology, 85*, 101-104.

Willerman, L. (1979). The effects of family on intellectual development. *American Psychologist, 34*, 923-929.

Witkin, H., & Berry, J. (1975). Psychological differentiation in cross-cultural perspective. *Journal of Cross Cultural Psychology, 6*, 4-87.

Wohlwill, J. (1973a). *The study of behavioral development*. New York: Academic Press.

Wohlwill, J. (1973b). The concept of experience: S or R. *Human Development, 16*, 90-107.

Wohlwill, J., & Heft, H. (1987). The physical environment and the development of the child. In I. Altman & J. D. Stokols (Eds.), *Handbook of environmental psychology*. New York: John Wiley.

Woodhead, M. (1988). When psychology informs public policy: The case of early childhood intervention. *American Psychologist, 43*, 443-454.

Woodson, R., & DaCosta-Woodson, I. (1984). Social organization, physical environment and infant caregiver interactions. *Developmental Psychology, 20*, 473-476.

Yarrow, L., & Anderson, B. (1979). Procedures for studying parent infant interaction: A critique. In E. Thoman (Ed.), *Origins of social responsiveness*. Hillsdale, NJ: Lawrence Erlbaum.

Yarrow, L., & Goodwin, M. (1965). Some conceptual issues in the study of mother infant interaction. *American Journal of Orthopsychiatry, 35*, 473-481.

Yarrow, L., Goodwin, M., Mannheimer, H., & Milowe, I. (1973). Infancy experiences and cognitive and personality development at 10 years. In J. Stone, H. Smith, & L. Murphy (Eds.), *The competent infant*. New York: Basic Books.

Yarrow, L., MacTurk, K., Vietze, P., McCarthy, M., Kline, R., & McQuiston, S. (1984). The developmental course of parental stimulation and its relationship to mastery motivation during infancy. *Developmental Psychology, 20*, 492-503.

Yarrow, L., Rubenstein, J., & Pedersen, F. (1975). *Infant and environment*. New York: John Wiley.

Yarrow, M., Campbell, J., & Burton, R. (1970). Recollections of childhood: A study of the retrospective method. *Monographs of the Society for Research in Child Development, 35*.

Yolton, J. (1977). *The Locke reader*. London: Cambridge University Press.

Zarb, J. (1978). Correlates of recidivism and social adjustment among training school delinquents. *Canadian Journal of Behavioral Science, 10*, 317-328.

Zaslow, M. (1989). Sex differences in children's response to parental divorce. *American Journal of Orthopsychiatry, 59*, 118-141.

Zaslow, M., & Hayes, C. (1986). Sex differences in children's response to psychosocial stress. In M. Lamb, A. Brown, & B. Rogoff (Eds.), *Advances in developmental psychology* (Vol. 4). Hillsdale, NJ: Lawrence Erlbaum.

Zegiob, L., Arnold, S., & Forehand, R. (1975). An examination of observer effects in parent child interactions. *Child Development, 46*, 509-512.

Zentall, S., & Shaw, J. (1980). Effects of classroom noise on performance and activity of 2nd grade hyperactive and control children. *Journal of Educational Psychology, 72*, 830-840.

Zentall, S., & Zentall, T. (1983). Optimal stimulation: A model of disordered activity in normal and deviant children. *Psychological Bulletin, 94*, 446-471.

Zigler, E., & Balla, D. (1982). Motivational and personality factors in the performance of the retarded. In E. Zigler & D. Balla (Eds.), *Mental retardation.* Hillsdale, NJ: Lawrence Erlbaum.

Zimmerman, D., & Williams, R. (1986). Note on the reliability of experimental measures and the power of significance tests. *Psychological Bulletin, 100*, 123-124.

Index

Age Specificity Hypothesis, 65-69
Attachment history:
   and environment, 81, 102-103
   as a moderator, 105

Child report measure, 17-18
Chronosystem, 43, 72, 74, 101, 106, 150
Cognitive status and environment, 103
Control systems theory, 104-106
Cross-cultural studies, 11, 15-16, 28-32,
   50-51, 53-54, 85-86, 99, 108-109

Day care, 33-34

Environmental influences:
   and cognitive development, 27-28, 31-
      32, 34-36
   and social emotional development,
      29-34, 36-37
   early history of, 2-5
   early versus later, 69-65
   experimental studies of, 18-19, 24-25
   modern history of, 6-9

sensitization versus steeling of children, 123-126
Environment as a system, 39, 106
   implications for environmental the-
      ory, 149-153
   implications for future research, 144-
      149, 157-158
   implications for intervention, 153-159
   nature of, 141-143
Exosystem, 41-42
   mediation, 57
   relation to caregiver behavior, 48
   relation to caregiver values, 45

Goodness of fit model, 109-111

Intervention studies, 14-15

Longitudinal studies, 22-23

Macrosystem, 42, 49-51
   mediation, 52-54, 98, 101, 105, 109
   relation to caregiver behaviors, 45-47

relation to caregiver values, 44-45
Mesosystem, 40
    mediation, 54
    relation to caregiver behaviors, 48-49
Microsystem, 40, 45
    mediation, 57-58, 105

Nature-nurture designs, 12-15
Noise-crowding influences, 56-57, 126-
    127

Observational studies, 19-21, 23-24
Organism-environment covariance:
    active, 93, 96-97, 103, 133, 155
    and causality, 111-112
    and interaction, 118, 133, 138-139
    biomedical-environment, 97-98
    environment-environment, 94-96
    gene-environment, 96-97
    nutrition-environment, 99
    passive, 93-94, 96, 99
    probabilistic versus deterministic mod-
        els, 93-95
    reactive, 93, 96-97, 99, 103-104
    risk status, 107-109
    temperament-environment, 100-102
Organism-environment interaction, 144,
    155-156
    biological risk x environment, 119-
        122
    buffering interactions, 121-123, 152,
        156-157

child behavior x environment, 132-
    133
clinical studies, 114-115
higher order interactions, 119, 134-
    135, 137-138, 147, 152
history, 113-114
in nonhuman species, 114
inconsistent findings, 115-119
nature of, 137-139
sex differences, 126-128, 134-137
temperament x environment, 128-132

Parent report measures, 16-17, 21
Predeterminism, 1-2, 4-5

Resilient children, 110-111, 156-157

Schools, 31-32
Sensitive periods, 60-66
Social address model, 6, 11-12, 21-22
Social support, 48-49, 51, 101
Specificity model of environmental ac-
    tion, 76-78, 86-87
    current status, 78-82, 82-86
    implications for research design, 90-
        91
    methodological and statistical criteria,
        82-82
    relation to model of nonshared envi-
        ronment, 88-90
Statistical power, 19, 54, 118, 148-149
Structure of the environment, 42-44, 150

About the Author

Theodore D. Wachs is Professor of Psychological Science at Purdue University. He received his Ph.D. in 1967 from George Peabody College, majoring in child clinical psychology. He is coauthor of *Early Experience and Human Development* (with G. Gruen) and coeditor of *Assessment of Young Developmentally Disabled Children* (with R. Sheehan) and *Conceptualization and Measurement of Organism-Environment Interaction* (with R. Plomin). He is a member of the editorial boards of *Child Development* and the *International Journal of Behavioral Development* as well as a fellow of the American Psychological Association. His current research interests involve investigation of the influence on development of the interaction between young children's nutritional status and their rearing environment as well as studies on the relation of noise and crowding to early development.